Practice

Eureka Math®
Grade 2 Fluency
Modules 1–5

Published by Great Minds®.

Copyright © 2015 Great Minds®. No part of this work may be reproduced, sold, or commercialized, in whole or in part, without written permission from Great Minds®. Noncommercial use is licensed pursuant to a Creative Commons Attribution-NonCommercial-ShareAlike 4.0 license; for more information, go to http://greatminds.org/copyright. *Great Minds* and *Eureka Math* are registered trademarks of Great Minds®.

Printed in the U.S.A.

This book may be purchased from the publisher at eureka-math.org.

BAB   10 9 8 7 6 5 4 3 2 1

ISBN 978-1-64054-580-9

G2-M1-M5-P/F-04.2018

Learn ◆ Practice ◆ Succeed

Eureka Math® student materials for *A Story of Units*® (K–5) are available in the *Learn, Practice, Succeed* trio. This series supports differentiation and remediation while keeping student materials organized and accessible. Educators will find that the *Learn, Practice,* and *Succeed* series also offers coherent—and therefore, more effective—resources for Response to Intervention (RTI), extra practice, and summer learning.

Learn

Eureka Math Learn serves as a student's in-class companion where they show their thinking, share what they know, and watch their knowledge build every day. *Learn* assembles the daily classwork—Application Problems, Exit Tickets, Problem Sets, templates—in an easily stored and navigated volume.

Practice

Each *Eureka Math* lesson begins with a series of energetic, joyous fluency activities, including those found in *Eureka Math Practice.* Students who are fluent in their math facts can master more material more deeply. With *Practice,* students build competence in newly acquired skills and reinforce previous learning in preparation for the next lesson.

Together, *Learn* and *Practice* provide all the print materials students will use for their core math instruction.

Succeed

Eureka Math Succeed enables students to work individually toward mastery. These additional problem sets align lesson by lesson with classroom instruction, making them ideal for use as homework or extra practice. Each problem set is accompanied by a Homework Helper, a set of worked examples that illustrate how to solve similar problems.

Teachers and tutors can use *Succeed* books from prior grade levels as curriculum-consistent tools for filling gaps in foundational knowledge. Students will thrive and progress more quickly as familiar models facilitate connections to their current grade-level content.

Students, families, and educators:

Thank you for being part of the *Eureka Math®* community, where we celebrate the joy, wonder, and thrill of mathematics. One of the most obvious ways we display our excitement is through the fluency activities provided in *Eureka Math Practice*.

What is fluency in mathematics?

You may think of *fluency* as associated with the language arts, where it refers to speaking and writing with ease. In prekindergarten through grade 5, the *Eureka Math* curriculum contains multiple daily opportunities to build fluency *in mathematics*. Each is designed with the same notion—growing every student's ability to use mathematics *with ease*. Fluency experiences are generally fast-paced and energetic, celebrating improvement and focusing on recognizing patterns and connections within the material. They are not intended to be graded.

Eureka Math fluency activities provide differentiated practice through a variety of formats—some are conducted orally, some use manipulatives, others use a personal whiteboard, and still others use a handout and paper-and-pencil format. *Eureka Math Practice* provides each student with the printed fluency exercises for his or her grade level.

What is a Sprint?

Many printed fluency activities utilize the format we call a Sprint. These exercises build speed and accuracy with already acquired skills. Used when students are nearing optimum proficiency, Sprints leverage tempo to build a low-stakes adrenaline boost that increases memory and recall. Their intentional design makes Sprints inherently differentiated; the problems build from simple to complex, with the first quadrant of problems being the simplest and each subsequent quadrant adding complexity. Further, intentional patterns within the sequence of problems engage students' higher order thinking skills.

The suggested format for delivering a Sprint calls for students to do two consecutive Sprints (labeled A and B) on the same skill, each timed at one minute. Students pause between Sprints to articulate the patterns they noticed as they worked the first Sprint. Noticing the patterns often provides a natural boost to their performance on the second Sprint.

Sprints can be conducted with an untimed protocol as well. The untimed protocol is highly recommended when students are still building confidence with the level of complexity of the first quadrant of problems. Once all students are prepared for success on the Sprint, the work of improving speed and accuracy with the energy of a timed protocol is often welcome and invigorating.

Where can I find other fluency activities?

The *Eureka Math Teacher Edition* guides educators in the delivery of all fluency activities for each lesson, including those that do not require print materials. Additionally, the *Eureka Digital Suite* provides access to the fluency activities for all grade levels, searchable by standard or lesson.

Best wishes for a year filled with aha moments!

Jill Diniz

Jill Diniz
Director of Mathematics
Great Minds

Contents

Module 1

Module 2

Module 3

Module 4

Module 5

Grade 2

Module 1

A

Number Correct: _____

Name _____ Date _____

Add a Ten and Some Ones

1.	$10 + 1 = \underline{\hspace{1cm}}$	16.	$3 + 10 = \underline{\hspace{1cm}}$
2.	$10 + 2 = \underline{\hspace{1cm}}$	17.	$4 + 10 = \underline{\hspace{1cm}}$
3.	$10 + 4 = \underline{\hspace{1cm}}$	18.	$1 + 10 = \underline{\hspace{1cm}}$
4.	$10 + 3 = \underline{\hspace{1cm}}$	19.	$2 + 10 = \underline{\hspace{1cm}}$
5.	$10 + 5 = \underline{\hspace{1cm}}$	20.	$5 + 10 = \underline{\hspace{1cm}}$
6.	$10 + 6 = \underline{\hspace{1cm}}$	21.	$\underline{\hspace{1cm}} = 10 + 5$
7.	$\underline{\hspace{1cm}} = 10 + 1$	22.	$\underline{\hspace{1cm}} = 10 + 8$
8.	$\underline{\hspace{1cm}} = 10 + 4$	23.	$\underline{\hspace{1cm}} = 10 + 9$
9.	$\underline{\hspace{1cm}} = 10 + 3$	24.	$\underline{\hspace{1cm}} = 10 + 6$
10.	$\underline{\hspace{1cm}} = 10 + 5$	25.	$\underline{\hspace{1cm}} = 10 + 7$
11.	$\underline{\hspace{1cm}} = 10 + 2$	26.	$16 = \underline{\hspace{1cm}} + 6$
12.	$10 + 6 = \underline{\hspace{1cm}}$	27.	$8 + \underline{\hspace{1cm}} = 18$
13.	$10 + 9 = \underline{\hspace{1cm}}$	28.	$\underline{\hspace{1cm}} + 10 = 17$
14.	$10 + 7 = \underline{\hspace{1cm}}$	29.	$19 = \underline{\hspace{1cm}} + 10$
15.	$10 + 8 = \underline{\hspace{1cm}}$	30.	$18 = 8 + \underline{\hspace{1cm}}$

B

Improvement: _____   Number Correct: _____

Name _____   Date _____

Add a Ten and Some Ones

1.	$10 + 5 =$ _____	16.	$4 + 10 =$ _____
2.	$10 + 4 =$ _____	17.	$3 + 10 =$ _____
3.	$10 + 3 =$ _____	18.	$2 + 10 =$ _____
4.	$10 + 2 =$ _____	19.	$1 + 10 =$ _____
5.	$10 + 1 =$ _____	20.	$3 + 10 =$ _____
6.	$10 + 5 =$ _____	21.	_____ $= 10 + 6$
7.	_____ $= 10 + 4$	22.	_____ $= 10 + 9$
8.	_____ $= 10 + 2$	23.	_____ $= 10 + 5$
9.	_____ $= 10 + 1$	24.	_____ $= 10 + 7$
10.	_____ $= 10 + 3$	25.	_____ $= 10 + 8$
11.	_____ $= 10 + 4$	26.	$17 =$ _____ $+ 7$
12.	$10 + 6 =$ _____	27.	$3 +$ _____ $= 13$
13.	$10 + 7 =$ _____	28.	_____ $+ 10 = 16$
14.	$10 + 9 =$ _____	29.	$18 =$ _____ $+ 10$
15.	$10 + 8 =$ _____	30.	$17 = 7 +$ _____

Target Number:

Target Practice

Choose a *target number*, and write it in the middle of the circle on the top of the page. Roll a die. Write the number rolled in the circle at the end of one of the arrows. Then, make a bull's eye by writing the number needed to make your target in the other circle.

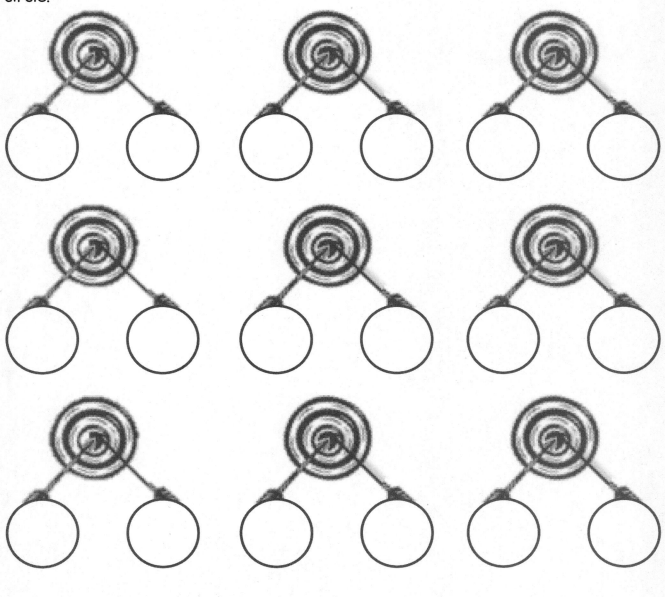

_____

target practice

EUREKA MATH   **Lesson 1:**   Practice making ten and adding to ten.

© 2015 Great Minds®. eureka-math.org

A

Number Correct: _____

Name _____ Date _____

Add Tens and Ones

1.	$10 + 3 = $ _____	16.	$10 + $ _____ $= 13$
2.	$20 + 2 = $ _____	17.	$40 + $ _____ $= 42$
3.	$30 + 4 = $ _____	18.	$60 + $ _____ $= 61$
4.	$50 + 3 = $ _____	19.	$70 + $ _____ $= 75$
5.	$20 + 5 = $ _____	20.	$80 + $ _____ $= 83$
6.	$50 + 5 = $ _____	21.	$60 + 9 = $ _____
7.	_____ $= 40 + 1$	22.	$80 + 9 = $ _____
8.	_____ $= 20 + 4$	23.	$80 + $ ___ $= 86$
9.	_____ $= 20 + 3$	24.	$90 + $ ___ $= 97$
10.	_____ $= 30 + 5$	25.	___ $+ 6 = 76$
11.	_____ $= 40 + 5$	26.	___ $+ 6 = 86$
12.	$30 + 6 = $ _____	27.	$86 = $ ___ $+ 6$
13.	$20 + 9 = $ _____	28.	___ $+ 60 = 67$
14.	$40 + 7 = $ _____	29.	$95 = $ ___ $+ 90$
15.	$50 + 8 = $ _____	30.	$97 = 7 + $ ___

B

Improvement: _____   Number Correct: _____

Name _____   Date _____

Add Tens and Ones

1.	$10 + 2 =$ _____	16.	$10 +$ _____ $= 12$
2.	$20 + 3 =$ _____	17.	$40 +$ _____ $= 42$
3.	$30 + 4 =$ _____	18.	$60 +$ _____ $= 61$
4.	$50 + 4 =$ _____	19.	$70 +$ _____ $= 75$
5.	$40 + 5 =$ _____	20.	$80 +$ _____ $= 83$
6.	$50 + 1 =$ _____	21.	$70 + 8 =$ _____
7.	_____ $= 50 + 1$	22.	$80 + 8 =$ _____
8.	_____ $= 20 + 4$	23.	$70 +$ _____ $= 76$
9.	_____ $= 20 + 2$	24.	$90 +$ _____ $= 99$
10.	_____ $= 30 + 5$	25.	_____ $+ 8 = 78$
11.	_____ $= 40 + 3$	26.	_____ $+ 6 = 96$
12.	$30 + 7 =$ _____	27.	$86 =$ _____ $+ 6$
13.	$20 + 8 =$ _____	28.	_____ $+ 60 = 67$
14.	$40 + 9 =$ _____	29.	$95 =$ _____ $+ 90$
15.	$50 + 6 =$ _____	30.	$97 = 7 +$ _____

A

Number Correct: _____

Name _____ Date _____

*Write the missing number. Pay attention to the + and – signs.

1.	$3 + 1 =$ ___	16.	$6 + 2 =$ ___
2.	$13 + 1 =$ ___	17.	$56 + 2 =$ ___
3.	$23 + 1 =$ ___	18.	$7 + 2 =$ ___
4.	$1 + 2 =$ ___	19.	$67 + 2 =$ ___
5.	$11 + 2 =$ ___	20.	$87 + 2 =$ ___
6.	$21 + 2 =$ ___	21.	$7 - 2 =$ ___
7.	$31 + 2 =$ ___	22.	$47 - 2 =$ ___
8.	$61 + 2 =$ ___	23.	$67 - 2 =$ ___
9.	$4 - 1 =$ ___	24.	$26 + 3 =$ ___
10.	$14 - 1 =$ ___	25.	$56 +$ ___ $= 59$
11.	$24 - 1 =$ ___	26.	___ $+ 3 = 76$
12.	$54 - 1 =$ ___	27.	$57 -$ ___ $= 54$
13.	$5 - 3 =$ ___	28.	$77 -$ ___ $= 74$
14.	$15 - 3 =$ ___	29.	___ $- 4 = 73$
15.	$25 - 3 =$ ___	30.	___ $- 4 = 93$

B

Improvement: _____   Number Correct: _____

Name _____   Date _____

*Write the missing number. Pay attention to the + and − signs.

1.	2 + 1 = ___	16.	7 + 2 = ___
2.	12 + 1 = ___	17.	67 + 2 = ___
3.	22 + 1 = ___	18.	4 + 5 = ___
4.	3 + 2 = ___	19.	54 + 5 = ___
5.	13 + 2 = ___	20.	84 + 5 = ___
6.	23 + 2 = ___	21.	8 − 6 = ___
7.	43 + 2 = ___	22.	48 − 6 = ___
8.	63 + 2 = ___	23.	78 − 6 = ___
9.	5 − 1 = ___	24.	33 + 4 = ___
10.	15 − 1 = ___	25.	63 + ___ = 67
11.	25 − 1 = ___	26.	___ + 3 = 77
12.	45 − 1 = ___	27.	59 − ___ = 56
13.	5 − 4 = ___	28.	79 − ___ = 76
14.	15 − 4 = ___	29.	___ − 6 = 73
15.	25 − 4 = ___	30.	___ − 6 = 93

Grade 2
Module 2

A

Number Correct: _____

Before, Between, After

1.	1, 2, ____	
2.	11, 12, ____	
3.	21, 22, ____	
4.	71, 72, ____	
5.	3, 4, ____	
6.	3, ____, 5	
7.	13, ____, 15	
8.	23, ____, 25	
9.	83, ____, 85	
10.	7, 8, ____	
11.	7, ____, 9	
12.	____, 8, 9	
13.	____, 18, 19	
14.	____, 28, 29	
15.	____, 58, 59	
16.	12, 13, ____	
17.	45, 46, ____	
18.	12, ____, 14	
19.	36, ____, 38	
20.	____, 19, 20	
21.	____, 89, 90	
22.	98, 99, ____	

23.	99, ____, 101	
24.	19, 20, ____	
25.	119, 120, ____	
26.	35, ____, 37	
27.	135, ____, 137	
28.	____, 24, 25	
29.	____, 124, 125	
30.	142, 143, ____	
31.	138, ____, 140	
32.	____, 149, 150	
33.	148, ____, 150	
34.	____, 149, 150	
35.	____, 163, 164	
36.	187, ____, 189	
37.	____, 170, 171	
38.	178, 179, ____	
39.	192, ____, 194	
40.	____, 190, 191	
41.	197, ____, 199	
42.	168, 169, ____	
43.	199, ____, 201	
44.	____, 160, 161	

EUREKA MATH®

Lesson 1: Connect measurement with physical units by using multiple copies of the same physical unit to measure.

© 2015 Great Minds®. eureka-math.org

19

B

Number Correct: _____

Improvement: _____

Before, Between, After

1.	0, 1, ____	
2.	10, 11, ____	
3.	20, 21, ____	
4.	70, 71, ____	
5.	2, 3, ____	
6.	2, ____, 4	
7.	12, ____, 14	
8.	22, ____ 24	
9.	82, ____, 84	
10.	6, 7, ____	
11.	6, ____, 8	
12.	____, 7, 8	
13.	____, 17, 18	
14.	____, 27, 28	
15.	____, 57, 58	
16.	11, 12, ____	
17.	44, 45, ____	
18.	11, ____, 13	
19.	35, ____, 37	
20.	____, 19, 20	
21.	____, 79, 80	
22.	98, 99, ____	

23.	99, ____, 101	
24.	29, 30, ____	
25.	129, 130, ____	
26.	34, ____, 36	
27.	134, ____, 136	
28.	____, 23, 24	
29.	____, 123, 124	
30.	141, 142, ____	
31.	128, ____, 130	
32.	____, 149, 150	
33.	148, ____, 150	
34.	____, 149, 150	
35.	____, 173, 174	
36.	167, ____, 169	
37.	____, 160, 161	
38.	188, 189, ____	
39.	193, ____, 195	
40.	____, 170, 171	
41.	196, ____, 198	
42.	178, 179, ____	
43.	199, ____, 201	
44.	____, 180, 181	

EUREKA MATH

Lesson 1:   Connect measurement with physical units by using multiple copies of
the same physical unit to measure.

© 2015 Great Minds®. eureka-math.org

21

A

Number Correct: _____

Making Ten

1.	0 + ___ = 10	
2.	9 + ___ = 10	
3.	8 + ___ = 10	
4.	7 + ___ = 10	
5.	6 + ___ = 10	
6.	5 + ___ = 10	
7.	1 + ___ = 10	
8.	2 + ___ = 10	
9.	3 + ___ = 10	
10.	4 + ___ = 10	
11.	10 + ___ = 10	
12.	9 + ___ = 10	
13.	19 + ___ = 20	
14.	5 + ___ = 10	
15.	15 + ___ = 20	
16.	8 + ___ = 10	
17.	18 + ___ = 20	
18.	6 + ___ = 10	
19.	16 + ___ = 20	
20.	7 + ___ = 10	
21.	17 + ___ = 20	
22.	3 + ___ = 10	

23.	13 + ___ = 20	
24.	23 + ___ = 30	
25.	27 + ___ = 30	
26.	5 + ___ = 10	
27.	25 + ___ = 30	
28.	2 + ___ = 10	
29.	22 + ___ = 30	
30.	32 + ___ = 40	
31.	1 + ___ = 10	
32.	11 + ___ = 20	
33.	21 + ___ = 30	
34.	31 + ___ = 40	
35.	38 + ___ = 40	
36.	36 + ___ = 40	
37.	39 + ___ = 40	
38.	35 + ___ = 40	
39.	___ + 6 = 30	
40.	___ + 8 = 20	
41.	___ + 7 = 40	
42.	___ + 6 = 20	
43.	___ + 4 = 30	
44.	___ + 8 = 40	

EUREKA MATH®

Lesson 3: Apply concepts to create unit rulers and measure lengths using unit rulers.

© 2015 Great Minds®. eureka-math.org

23

B

Number Correct: _____

Improvement: _____

Making Ten

1.	10 + ___ = 10	
2.	9 + ___ = 10	
3.	8 + ___ = 10	
4.	7 + ___ = 10	
5.	6 + ___ = 10	
6.	5 + ___ = 10	
7.	1 + ___ = 10	
8.	2 + ___ = 10	
9.	3 + ___ = 10	
10.	4 + ___ = 10	
11.	0 + ___ = 10	
12.	5 + ___ = 10	
13.	15 + ___ = 20	
14.	9 + ___ = 10	
15.	19 + ___ = 20	
16.	8 + ___ = 10	
17.	18 + ___ = 20	
18.	7 + ___ = 10	
19.	17 + ___ = 20	
20.	6 + ___ = 10	
21.	16 + ___ = 20	
22.	4 + ___ = 10	

23.	14 + ___ = 20	
24.	24 + ___ = 30	
25.	26 + ___ = 30	
26.	9 + ___ = 10	
27.	29 + ___ = 30	
28.	3 + ___ = 10	
29.	23 + ___ = 30	
30.	33 + ___ = 40	
31.	2 + ___ = 10	
32.	12 + ___ = 20	
33.	22 + ___ = 30	
34.	32 + ___ = 40	
35.	37 + ___ = 40	
36.	34 + ___ = 40	
37.	35 + ___ = 40	
38.	39 + ___ = 40	
39.	___ + 4 = 30	
40.	___ + 9 = 20	
41.	___ + 4 = 40	
42.	___ + 7 = 20	
43.	___ + 3 = 30	
44.	___ + 9 = 40	

EUREKA MATH

Lesson 3: Apply concepts to create unit rulers and measure lengths using unit rulers.

25

© 2015 Great Minds®. eureka-math.org

A

Number Correct: _____

Related Facts

1.	8 + 3 =		23.	15 – 6 =	
2.	3 + 8 =		24.	15 – 9 =	
3.	11 – 3 =		25.	8 + 7 =	
4.	11 – 8 =		26.	7 + 8 =	
5.	7 + 4 =		27.	15 – 7 =	
6.	4 + 7 =		28.	15 – 8 =	
7.	11 – 4 =		29.	9 + 4 =	
8.	11 – 7 =		30.	4 + 9 =	
9.	9 + 3 =		31.	13 – 4 =	
10.	3 + 9 =		32.	13 – 9 =	
11.	12 – 3 =		33.	8 + 6 =	
12.	12 – 9 =		34.	6 + 8 =	
13.	8 + 5 =		35.	14 – 6 =	
14.	5 + 8 =		36.	14 – 8 =	
15.	13 – 5 =		37.	7 + 6 =	
16.	13 – 8 =		38.	6 + 7 =	
17.	7 + 5 =		39.	13 – 6 =	
18.	5 + 7 =		40.	13 – 7 =	
19.	12 – 5 =		41.	9 + 7 =	
20.	12 – 7 =		42.	7 + 9 =	
21.	9 + 6 =		43.	16 – 7 =	
22.	6 + 9 =		44.	16 – 9 =	

Lesson 4: Measure various objects using centimeter rulers and meter sticks.

B

Related Facts

Number Correct: _____

Improvement: _____

1.	9 + 2 =	
2.	2 + 9 =	
3.	11 – 2 =	
4.	11 – 9 =	
5.	6 + 5 =	
6.	5 + 6 =	
7.	11 – 5 =	
8.	11 – 6 =	
9.	8 + 4 =	
10.	4 + 8 =	
11.	12 – 4 =	
12.	12 – 8 =	
13.	7 + 6 =	
14.	6 + 7 =	
15.	13 – 6 =	
16.	13 – 7 =	
17.	9 + 3 =	
18.	3 + 9 =	
19.	12 – 3 =	
20.	12 – 9 =	
21.	8 + 7 =	
22.	7 + 8 =	

23.	15 – 7 =	
24.	15 – 8 =	
25.	9 + 6 =	
26.	6 + 9 =	
27.	15 – 6 =	
28.	15 – 9 =	
29.	7 + 5 =	
30.	5 + 7 =	
31.	12 – 5 =	
32.	12 – 7 =	
33.	9 + 5 =	
34.	5 + 9 =	
35.	14 – 5 =	
36.	14 – 9 =	
37.	8 + 6 =	
38.	6 + 8 =	
39.	14 – 6 =	
40.	14 – 8 =	
41.	9 + 8 =	
42.	8 + 9 =	
43.	17 – 8 =	
44.	17 – 9 =	

Lesson 4:   Measure various objects using centimeter rulers and meter sticks.

29

A

Number Correct: _____

Circle the longer length.

1.	1 cm	0 cm		23.	110 cm	101 cm
2.	11 cm	10 cm		24.	110 cm	1 m
3.	11 cm	12 cm		25.	1 m	111 cm
4.	22 cm	12 cm		26.	101 cm	1 m
5.	29 cm	30 cm		27.	111 cm	101 cm
6.	31 cm	13 cm		28.	112 cm	102 cm
7.	43 cm	33 cm		29.	110 cm	115 cm
8.	33 cm	23 cm		30.	115 cm	105 cm
9.	35 cm	53 cm		31.	106 cm	116 cm
10.	50 cm	35 cm		32.	108 cm	98 cm
11.	55 cm	45 cm		33.	119 cm	99 cm
12.	50 cm	55 cm		34.	131 cm	133 cm
13.	65 cm	56 cm		35.	133 cm	113 cm
14.	66 cm	56 cm		36.	142 cm	124 cm
15.	66 cm	86 cm		37.	144 cm	114 cm
16.	86 cm	68 m		38.	154 cm	145 cm
17.	68 cm	88 cm		39.	155 cm	152 cm
18.	89 cm	98 cm		40.	198 cm	199 cm
19.	99 cm	98 m		41.	215 cm	225 cm
20.	99 cm	1 m		42.	252 cm	255 cm
21.	1 m	101 cm		43.	2 m	295 cm
22.	1 m	90 cm		44.	3 m	295 cm

EUREKA MATH®

Lesson 6:   Measure and compare lengths using centimeters and meters.

B

Circle the longer length.

Number Correct: _____

Improvement: _____

1.	0 cm	1 cm	23.	111 cm	101 cm
2.	10 cm	12 cm	24.	101 cm	110 cm
3.	12 cm	11 cm	25.	1 m	110 cm
4.	32 cm	13 cm	26.	111 cm	1 m
5.	39 cm	40 cm	27.	113 cm	117 cm
6.	41 cm	14 cm	28.	112 cm	111 cm
7.	44 cm	40 cm	29.	115 cm	105 cm
8.	44 cm	54 cm	30.	106 cm	116 cm
9.	55 cm	65 cm	31.	107 cm	117 cm
10.	60 cm	59 cm	32.	118 cm	108 cm
11.	65 cm	45 cm	33.	119 cm	120 cm
12.	70 cm	65 cm	34.	132 cm	123 cm
13.	75 cm	57 cm	35.	133 cm	132 cm
14.	77 cm	76 cm	36.	143 cm	134 cm
15.	87 cm	78 cm	37.	144 cm	114 cm
16.	79 cm	97 m	38.	154 cm	145 cm
17.	79 cm	88 cm	39.	155 cm	152 cm
18.	98 cm	97 cm	40.	195 cm	199 cm
19.	99 cm	1 m	41.	225 cm	152 cm
20.	99 cm	100 cm	42.	252 cm	255 cm
21.	101 cm	100 cm	43.	2 m	295 cm
22.	1 m	101 cm	44.	3 m	295 cm

A

Number Correct: _____

Subtraction

1.	3 – 1 =		23.	8 – 7 =	
2.	13 – 1 =		24.	18 – 7 =	
3.	23 – 1 =		25.	58 – 7 =	
4.	53 – 1 =		26.	62 – 2 =	
5.	4 – 2 =		27.	9 – 8 =	
6.	14 – 2 =		28.	19 – 8 =	
7.	24 – 2 =		29.	29 – 8 =	
8.	64 – 2 =		30.	69 – 8 =	
9.	4 – 3 =		31.	7 – 3 =	
10.	14 – 3 =		32.	17 – 3 =	
11.	24 – 3 =		33.	77 – 3 =	
12.	74 – 3 =		34.	59 – 9 =	
13.	6 – 4 =		35.	9 – 7 =	
14.	16 – 4 =		36.	19 – 7 =	
15.	26 – 4 =		37.	89 – 7 =	
16.	96 – 4 =		38.	99 – 5 =	
17.	7 – 5 =		39.	78 – 6 =	
18.	17 – 5 =		40.	58 – 5 =	
19.	27 – 5 =		41.	39 – 7 =	
20.	47 – 5 =		42.	28 – 6 =	
21.	43 – 3 =		43.	49 – 4 =	
22.	87 – 7 =		44.	67 – 4 =	

EUREKA MATH®

Lesson 7:   Measure and compare lengths using standard metric length units and non-standard length unitsÃrelate measurement to unit size.

35

© 2015 Great Minds®. eureka-math.org

B

Number Correct: _____

Improvement: _____

Subtraction

1.	2 – 1 =		23.	8 – 7 =	
2.	12 – 1 =		24.	18 – 7 =	
3.	22 – 1 =		25.	68 – 7 =	
4.	52 – 1 =		26.	32 – 2 =	
5.	5 – 2 =		27.	9 – 8 =	
6.	15 – 2 =		28.	19 – 8 =	
7.	25 – 2 =		29.	29 – 8 =	
8.	65 – 2 =		30.	79 – 8 =	
9.	4 – 3 =		31.	8 – 4 =	
10.	14 – 3 =		32.	18 – 4 =	
11.	24 – 3 =		33.	78 – 4 =	
12.	84 – 3 =		34.	89 – 9 =	
13.	7 – 4 =		35.	9 – 7 =	
14.	17 – 4 =		36.	19 – 7 =	
15.	27 – 4 =		37.	79 – 7 =	
16.	97 – 4 =		38.	89 – 5 =	
17.	6 – 5 =		39.	68 – 6 =	
18.	16 – 5 =		40.	48 – 5 =	
19.	26 – 5 =		41.	29 – 7 =	
20.	46 – 5 =		42.	38 – 6 =	
21.	23 – 3 =		43.	59 – 4 =	
22.	67 – 7 =		44.	77 – 4 =	

EUREKA MATH®

Lesson 7:   Measure and compare lengths using standard metric length units and non-standard length units—relate measurement to unit size.

37

A

Number Correct: _____

Making a Meter

1.	10 cm + ____ = 100 cm	
2.	30 cm + ____ = 100 cm	
3.	50 cm + ____ = 100 cm	
4.	70 cm + ____ = 100 cm	
5.	90 cm + ____ = 100 cm	
6.	80 cm + ____ = 100 cm	
7.	60 cm + ____ = 100 cm	
8.	40 cm + ____ = 100 cm	
9.	20 cm + ____ = 100 cm	
10.	21 cm + ____ = 100 cm	
11.	23 cm + ____ = 100 cm	
12.	25 cm + ____ = 100 cm	
13.	27 cm + ____ = 100 cm	
14.	37 cm + ____ = 100 cm	
15.	38 cm + ____ = 100 cm	
16.	39 cm + ____ = 100 cm	
17.	49 cm + ____ = 100 cm	
18.	50 cm + ____ = 100 cm	
19.	52 cm + ____ = 100 cm	
20.	56 cm + ____ = 100 cm	
21.	58 cm + ____ = 100 cm	
22.	62 cm + ____ = 100 cm	

23.	____ + 62 cm = 1 m	
24.	____ + 72 cm = 1 m	
25.	____ + 92 cm = 1 m	
26.	____ + 29 cm = 1 m	
27.	____ + 39 cm = 1 m	
28.	____ + 59 cm = 1 m	
29.	____ + 89 cm = 1 m	
30.	____ + 88 cm = 1 m	
31.	____ + 68 cm = 1 m	
32.	____ + 18 cm = 1 m	
33.	____ + 15 cm = 1 m	
34.	____ + 55 cm = 1 m	
35.	44 cm + ____ = 1 m	
36.	55 cm + ____ = 1 m	
37.	88 cm + ____ = 1 m	
38.	1 m = ____ + 33 cm	
39.	1 m = ____ + 66 cm	
40.	1 m = ____ + 99 cm	
41.	1 m – 11 cm = ____	
42.	1 m – 15 cm = ____	
43.	1 m – 17 cm = ____	
44.	1 m – 19 cm = ____	

EUREKA MATH®

Lesson 8: Solve addition and subtraction word problems using the ruler as a number line.

© 2015 Great Minds®. eureka-math.org

39

B

Number Correct: _____

Improvement: _____

Making a Meter

1.	1 cm + ____ = 100 cm	
2.	10 cm + ____ = 100 cm	
3.	20 cm + ____ = 100 cm	
4.	40 cm + ____ = 100 cm	
5.	60 cm + ____ = 100 cm	
6.	80 cm + ____ = 100 cm	
7.	90 cm + ____ = 100 cm	
8.	70 cm + ____ = 100 cm	
9.	50 cm + ____ = 100 cm	
10.	30 cm + ____ = 100 cm	
11.	31 cm + ____ = 100 cm	
12.	33 cm + ____ = 100 cm	
13.	35 cm + ____ = 100 cm	
14.	37 cm + ____ = 100 cm	
15.	39 cm + ____ = 100 cm	
16.	49 cm + ____ = 100 cm	
17.	59 cm + ____ = 100 cm	
18.	60 cm + ____ = 100 cm	
19.	62 cm + ____ = 100 cm	
20.	66 cm + ____ = 100 cm	
21.	68 cm + ____ = 100 cm	
22.	72 cm + ____ = 100 cm	

23.	____ + 72 cm = 1 m	
24.	____ + 82 cm = 1 m	
25.	____ + 28 cm = 1 m	
26.	____ + 38 cm = 1 m	
27.	____ + 48 cm = 1 m	
28.	____ + 45 cm = 1 m	
29.	____ + 43 cm = 1 m	
30.	____ + 34 cm = 1 m	
31.	____ + 24 cm = 1 m	
32.	____ + 14 cm = 1 m	
33.	____ + 12 cm = 1 m	
34.	____ + 10 cm = 1 m	
35.	11 cm + ____ = 1 m	
36.	33 cm + ____ = 1 m	
37.	55 cm + ____ = 1 m	
38.	1 m = ____ + 22 cm	
39.	1 m = ____ + 88 cm	
40.	1 m = ____ + 99 cm	
41.	1 m – 1 cm = ____	
42.	1 m – 5 cm = ____	
43.	1 m – 7 cm = ____	
44.	1 m – 17 cm = ____	

Lesson 8:   Solve addition and subtraction word problems using the ruler as a number line.

41

© 2015 Great Minds®. eureka-math.org

Grade 2

Module 3

A

Number Correct: _____

Differences to 10 with Teen Numbers

1.	3 – 1 =	
2.	13 – 1 =	
3.	5 – 1 =	
4.	15 – 1 =	
5.	7 – 1 =	
6.	17 – 1 =	
7.	4 – 2 =	
8.	14 – 2 =	
9.	6 – 2 =	
10.	16 – 2 =	
11.	8 – 2 =	
12.	18 – 2 =	
13.	4 – 3 =	
14.	14 – 3 =	
15.	6 – 3 =	
16.	16 – 3 =	
17.	8 – 3 =	
18.	18 – 3 =	
19.	6 – 4 =	
20.	16 – 4 =	
21.	8 – 4 =	
22.	18 – 4 =	

23.	7 – 4 =	
24.	17 – 4 =	
25.	7 – 5 =	
26.	17 – 5 =	
27.	9 – 5 =	
28.	19 – 5 =	
29.	7 – 6 =	
30.	17 – 6 =	
31.	9 – 6 =	
32.	19 – 6 =	
33.	8 – 7 =	
34.	18 – 7 =	
35.	9 – 8 =	
36.	19 – 8 =	
37.	7 – 3 =	
38.	17 – 3 =	
39.	5 – 4 =	
40.	15 – 4 =	
41.	8 – 5 =	
42.	18 – 5 =	
43.	8 – 6 =	
44.	18 – 6 =	

EUREKA MATH

Lesson 3: Count up and down between 90 and 1,000 using ones, tens, and hundreds.

45

B

Differences to 10 with Teen Numbers

Number Correct: _____

Improvement: _____

1.	2 – 1 =	
2.	12 – 1 =	
3.	4 – 1 =	
4.	14 – 1 =	
5.	6 – 1 =	
6.	16 – 1 =	
7.	3 – 2 =	
8.	13 – 2 =	
9.	5 – 2 =	
10.	15 – 2 =	
11.	7 – 2 =	
12.	17 – 2 =	
13.	5 – 3 =	
14.	15 – 3 =	
15.	7 – 3 =	
16.	17 – 3 =	
17.	9 – 3 =	
18.	19 – 3 =	
19.	5 – 4 =	
20.	15 – 4 =	
21.	7 – 4 =	
22.	17 – 4 =	

23.	9 – 4 =	
24.	19 – 4 =	
25.	6 – 5 =	
26.	16 – 5 =	
27.	8 – 5 =	
28.	18 – 5 =	
29.	8 – 6 =	
30.	18 – 6 =	
31.	9 – 6 =	
32.	19 – 6 =	
33.	9 – 7 =	
34.	19 – 7 =	
35.	9 – 8 =	
36.	19 – 8 =	
37.	8 – 3 =	
38.	18 – 3 =	
39.	6 – 4 =	
40.	16 – 4 =	
41.	9 – 5 =	
42.	19 – 5 =	
43.	7 – 6 =	
44.	17 – 6 =	

EUREKA MATH®

Lesson 3: Count up and down between 90 and 1,000 using ones, tens, and hundreds.

© 2015 Great Minds®. eureka-math.org

47

A

Number Correct: _____

Adding to the Teens

1.	5 + 5 + 5 =	
2.	9 + 1 + 3 =	
3.	2 + 8 + 4 =	
4.	3 + 7 + 2 =	
5.	4 + 6 + 9 =	
6.	9 + 0 + 6 =	
7.	3 + 0 + 8 =	
8.	2 + 7 + 7 =	
9.	6 + 6 + 6 =	
10.	7 + 8 + 4 =	
11.	3 + 5 + 9 =	
12.	9 + 1 + 1 =	
13.	5 + 5 + 6 =	
14.	8 + 2 + 8 =	
15.	3 + 4 + 7 =	
16.	5 + 0 + 8 =	
17.	6 + 2 + 6 =	
18.	6 + 3 + 9 =	
19.	2 + 4 + 7 =	
20.	3 + 8 + 6 =	
21.	5 + 7 + 6 =	
22.	3 + 6 + 9 =	

23.	1 + 9 + 5 =	
24.	3 + 5 + 5 =	
25.	8 + 4 + 6 =	
26.	9 + 7 + 1 =	
27.	2 + 6 + 8 =	
28.	0 + 8 + 7 =	
29.	8 + 4 + 3 =	
30.	9 + 2 + 2 =	
31.	4 + 4 + 4 =	
32.	6 + 8 + 5 =	
33.	4 + 5 + 7 =	
34.	7 + 3 + 1 =	
35.	6 + 4 + 3 =	
36.	1 + 9 + 9 =	
37.	5 + 8 + 5 =	
38.	3 + 3 + 5 =	
39.	7 + 0 + 6 =	
40.	4 + 5 + 9 =	
41.	4 + 8 + 4 =	
42.	2 + 6 + 7 =	
43.	3 + 5 + 6 =	
44.	2 + 6 + 9 =	

B

Number Correct: _____

Improvement: _____

Adding to the Teens

1.	5 + 5 + 4 =	
2.	7 + 3 + 5 =	
3.	1 + 9 + 8 =	
4.	4 + 6 + 2 =	
5.	2 + 8 + 9 =	
6.	7 + 0 + 6 =	
7.	4 + 0 + 9 =	
8.	2 + 9 + 9 =	
9.	4 + 5 + 4 =	
10.	8 + 7 + 5 =	
11.	2 + 7 + 9 =	
12.	9 + 1 + 2 =	
13.	6 + 4 + 5 =	
14.	8 + 2 + 3 =	
15.	1 + 4 + 9 =	
16.	3 + 8 + 0 =	
17.	7 + 4 + 7 =	
18.	5 + 3 + 8 =	
19.	7 + 3 + 4 =	
20.	5 + 8 + 6 =	
21.	7 + 6 + 4 =	
22.	5 + 8 + 4 =	

23.	8 + 2 + 5 =	
24.	9 + 1 + 6 =	
25.	3 + 6 + 4 =	
26.	3 + 2 + 7 =	
27.	4 + 8 + 6 =	
28.	9 + 9 + 0 =	
29.	0 + 7 + 5 =	
30.	8 + 4 + 4 =	
31.	3 + 8 + 8 =	
32.	5 + 7 + 6 =	
33.	3 + 4 + 9 =	
34.	3 + 7 + 3 =	
35.	6 + 4 + 5 =	
36.	7 + 9 + 1 =	
37.	2 + 6 + 8 =	
38.	5 + 3 + 7 =	
39.	6 + 0 + 9 =	
40.	2 + 5 + 7 =	
41.	3 + 6 + 3 =	
42.	4 + 2 + 9 =	
43.	6 + 3 + 5 =	
44.	7 + 2 + 9 =	

A

Number Correct: _____

Expanded Form

1.	20 + 1 =	
2.	20 + 2 =	
3.	20 + 3 =	
4.	20 + 9 =	
5.	30 + 9 =	
6.	40 + 9 =	
7.	80 + 9 =	
8.	40 + 4 =	
9.	50 + 5 =	
10.	10 + 7 =	
11.	20 + 5 =	
12.	200 + 30 =	
13.	300 + 40 =	
14.	400 + 50 =	
15.	500 + 60 =	
16.	600 + 70 =	
17.	700 + 80 =	
18.	200 + 30 + 5 =	
19.	300 + 40 + 5 =	
20.	400 + 50 + 6 =	
21.	500 + 60 + 7 =	
22.	600 + 70 + 8 =	

23.	400 + 20 + 5 =	
24.	200 + 60 + 1 =	
25.	200 + 1 =	
26.	300 + 1 =	
27.	400 + 1 =	
28.	500 + 1 =	
29.	700 + 1 =	
30.	300 + 50 + 2 =	
31.	300 + 2 =	
32.	100 + 10 + 7 =	
33.	100 + 7 =	
34.	700 + 10 + 5 =	
35.	700 + 5 =	
36.	300 + 40 + 7 =	
37.	300 + 7 =	
38.	500 + 30 + 2 =	
39.	500 + 2 =	
40.	2 + 500 =	
41.	2 + 600 =	
42.	2 + 40 + 600 =	
43.	3 + 10 + 700 =	
44.	8 + 30 + 700 =	

EUREKA MATH®

Lesson 7:   Write, read, and relate base ten numbers in all forms.

53

B

Expanded Form

Number Correct: _____

Improvement: _____

1.	10 + 1 =	
2.	10 + 2 =	
3.	10 + 3 =	
4.	10 + 9 =	
5.	20 + 9 =	
6.	30 + 9 =	
7.	70 + 9 =	
8.	30 + 3 =	
9.	40 + 4 =	
10.	80 + 7 =	
11.	90 + 5 =	
12.	100 + 20 =	
13.	200 + 30 =	
14.	300 + 40 =	
15.	400 + 50 =	
16.	500 + 60 =	
17.	600 + 70 =	
18.	300 + 40 + 5 =	
19.	400 + 50 + 6 =	
20.	500 + 60 + 7 =	
21.	600 + 70 + 8 =	
22.	700 + 80 + 9 =	

23.	500 + 30 + 6 =	
24.	300 + 70 + 1 =	
25.	300 + 1 =	
26.	400 + 1 =	
27.	500 + 1 =	
28.	600 + 1 =	
29.	900 + 1 =	
30.	400 + 60 + 3 =	
31.	400 + 3 =	
32.	100 + 10 + 5 =	
33.	100 + 5 =	
34.	800 + 10 + 5 =	
35.	800 + 5 =	
36.	200 + 30 + 7 =	
37.	200 + 7 =	
38.	600 + 40 + 2 =	
39.	600 + 2 =	
40.	2 + 600 =	
41.	3 + 600 =	
42.	3 + 40 + 600 =	
43.	5 + 10 + 800 =	
44.	9 + 20 + 700 =	

EUREKA MATH®

Lesson 7:   Write, read, and relate base ten numbers in all forms.

55

A

Number Correct: _____

Expanded Form

1.	100 + 20 + 3 =	
2.	100 + 20 + 4 =	
3.	100 + 20 + 5 =	
4.	100 + 20 + 8 =	
5.	100 + 30 + 8 =	
6.	100 + 40 + 8 =	
7.	100 + 70 + 8 =	
8.	500 + 10 + 9 =	
9.	500 + 10 + 8 =	
10.	500 + 10 + 7 =	
11.	500 + 10 + 3 =	
12.	700 + 30 =	
13.	700 + 3 =	
14.	30 + 3 =	
15.	700 + 33 =	
16.	900 + 40 =	
17.	900 + 4 =	
18.	40 + 4 =	
19.	900 + 44 =	
20.	800 + 70 =	
21.	800 + 7 =	
22.	70 + 7 =	

23.	800 + 77 =	
24.	300 + 90 + 2 =	
25.	400 + 80 =	
26.	600 + 7 =	
27.	200 + 60 + 4 =	
28.	100 + 9 =	
29.	500 + 80 =	
30.	80 + 500 =	
31.	2 + 50 + 400 =	
32.	2 + 400 + 50 =	
33.	3 + 70 + 800 =	
34.	40 + 9 + 800 =	
35.	700 + 9 + 20 =	
36.	5 + 300 =	
37.	400 + 90 + 10 =	
38.	500 + 80 + 20 =	
39.	900 + 60 + 40 =	
40.	400 + 80 + 2 =	
41.	300 + 60 + 5 =	
42.	200 + 27 + 5 =	
43.	8 + 700 + 59 =	
44.	47 + 500 + 8 =	

EUREKA MATH®

Lesson 10: Explore $1,000. How many $10 bills can we change for a thousand dollar bill?

57

B

Expanded Form

Number Correct: _____

Improvement: _____

1.	100 + 30 + 4 =		23.	700 + 66 =	
2.	100 + 30 + 5 =		24.	200 + 90 + 4 =	
3.	100 + 30 + 6 =		25.	500 + 70 =	
4.	100 + 30 + 9 =		26.	800 + 6 =	
5.	100 + 40 + 9 =		27.	400 + 70 + 4 =	
6.	100 + 50 + 9 =		28.	700 + 9 =	
7.	100 + 80 + 9 =		29.	800 + 50 =	
8.	400 + 10 + 8 =		30.	50 + 800 =	
9.	400 + 10 + 7 =		31.	2 + 80 + 400 =	
10.	400 + 10 + 6 =		32.	2 + 400 + 80 =	
11.	400 + 10 + 2 =		33.	3 + 70 + 500 =	
12.	700 + 80 =		34.	60 + 3 + 800 =	
13.	700 + 8 =		35.	900 + 7 + 20 =	
14.	80 + 8 =		36.	4 + 300 =	
15.	700 + 88 =		37.	500 + 90 + 10 =	
16.	900 + 20 =		38.	600 + 80 + 20 =	
17.	900 + 2 =		39.	900 + 60 + 40 =	
18.	20 + 2 =		40.	600 + 8 + 2 =	
19.	900 + 22 =		41.	800 + 6 + 5 =	
20.	700 + 60 =		42.	800 + 27 + 5 =	
21.	700 + 6 =		43.	8 + 100 + 49 =	
22.	60 + 6 =		44.	37 + 600 + 8 =	

A

Number Correct: _____

Addition and Subtraction to 10

1.	2 + 1 =	
2.	1 + 2 =	
3.	3 – 1 =	
4.	3 – 2 =	
5.	4 + 1 =	
6.	1 + 4 =	
7.	5 – 1 =	
8.	5 – 4 =	
9.	8 + 1 =	
10.	1 + 8 =	
11.	9 – 1 =	
12.	9 – 8 =	
13.	3 + 2 =	
14.	2 + 3 =	
15.	5 – 2 =	
16.	5 – 3 =	
17.	5 + 2 =	
18.	2 + 5 =	
19.	7 – 2 =	
20.	7 – 5 =	
21.	6 + 2 =	
22.	2 + 6 =	

23.	8 – 2 =	
24.	8 – 6 =	
25.	8 + 2 =	
26.	2 + 8 =	
27.	10 – 2 =	
28.	10 – 8 =	
29.	4 + 3 =	
30.	3 + 4 =	
31.	7 – 3 =	
32.	7 – 4 =	
33.	5 + 3 =	
34.	3 + 5 =	
35.	8 – 3 =	
36.	8 – 5 =	
37.	6 + 3 =	
38.	3 + 6 =	
39.	9 – 3 =	
40.	9 – 6 =	
41.	5 + 4 =	
42.	4 + 5 =	
43.	9 – 5 =	
44.	9 – 4 =	

Lesson 11: Count the total value of ones, tens, and hundreds with place value disks.

B

Number Correct: _____

Improvement: _____

Addition and Subtraction to 10

1.	3 + 1 =	
2.	1 + 3 =	
3.	4 – 1 =	
4.	4 – 3 =	
5.	5 + 1 =	
6.	1 + 5 =	
7.	6 – 1 =	
8.	6 – 5 =	
9.	9 + 1 =	
10.	1 + 9 =	
11.	10 – 1 =	
12.	10 – 9 =	
13.	4 + 2 =	
14.	2 + 4 =	
15.	6 – 2 =	
16.	6 – 4 =	
17.	7 + 2 =	
18.	2 + 7 =	
19.	9 – 2 =	
20.	9 – 7 =	
21.	5 + 2 =	
22.	2 + 5 =	

23.	7 – 2 =	
24.	7 – 5 =	
25.	8 + 2 =	
26.	2 + 8 =	
27.	10 – 2 =	
28.	10 – 8 =	
29.	4 + 3 =	
30.	3 + 4 =	
31.	7 – 3 =	
32.	7 – 4 =	
33.	5 + 3 =	
34.	3 + 5 =	
35.	8 – 3 =	
36.	8 – 5 =	
37.	7 + 3 =	
38.	3 + 7 =	
39.	10 – 3 =	
40.	10 – 7 =	
41.	5 + 4 =	
42.	4 + 5 =	
43.	9 – 5 =	
44.	9 – 4 =	

EUREKA MATH

Lesson 11: Count the total value of ones, tens, and hundreds with place value disks.

63

© 2015 Great Minds®. eureka-math.org

A

Number Correct: _____

Sums to 10 with Teen Numbers

1.	3 + 1 =	
2.	13 + 1 =	
3.	5 + 1 =	
4.	15 + 1 =	
5.	7 + 1 =	
6.	17 + 1 =	
7.	4 + 2 =	
8.	14 + 2 =	
9.	6 + 2 =	
10.	16 + 2 =	
11.	8 + 2 =	
12.	18 + 2 =	
13.	4 + 3 =	
14.	14 + 3 =	
15.	6 + 3 =	
16.	16 + 3 =	
17.	5 + 5 =	
18.	15 + 5 =	
19.	7 + 3 =	
20.	17 + 3 =	
21.	6 + 4 =	
22.	16 + 4 =	

23.	4 + 5 =	
24.	14 + 5 =	
25.	2 + 5 =	
26.	12 + 5 =	
27.	5 + 4 =	
28.	15 + 4 =	
29.	3 + 4 =	
30.	13 + 4 =	
31.	3 + 6 =	
32.	13 + 6 =	
33.	7 + 1 =	
34.	17 + 1 =	
35.	8 + 1 =	
36.	18 + 1 =	
37.	4 + 3 =	
38.	14 + 3 =	
39.	4 + 1 =	
40.	14 + 1 =	
41.	5 + 3 =	
42.	15 + 3 =	
43.	4 + 4 =	
44.	14 + 4 =	

EUREKA MATH®

Lesson 12: Change 10 ones for 1 ten, 10 tens for 1 hundred, and 10 hundreds for 1 thousand.

65

© 2015 Great Minds®. eureka-math.org

B

Number Correct: _____

Improvement: _____

Sums to 10 with Teen Numbers

1.	2 + 1 =	
2.	12 + 1 =	
3.	4 + 1 =	
4.	14 + 1 =	
5.	6 + 1 =	
6.	16 + 1 =	
7.	3 + 2 =	
8.	13 + 2 =	
9.	5 + 2 =	
10.	15 + 2 =	
11.	7 + 2 =	
12.	17 + 2 =	
13.	5 + 3 =	
14.	15 + 3 =	
15.	7 + 3 =	
16.	17 + 3 =	
17.	6 + 3 =	
18.	16 + 3 =	
19.	5 + 4 =	
20.	15 + 4 =	
21.	1 + 9 =	
22.	11 + 9 =	

23.	9 + 1 =	
24.	19 + 1 =	
25.	5 + 1 =	
26.	15 + 1 =	
27.	5 + 3 =	
28.	15 + 3 =	
29.	6 + 2 =	
30.	16 + 2 =	
31.	3 + 6 =	
32.	13 + 6 =	
33.	7 + 2 =	
34.	17 + 2 =	
35.	1 + 8 =	
36.	11 + 8 =	
37.	3 + 5 =	
38.	13 + 5 =	
39.	4 + 2 =	
40.	14 + 2 =	
41.	5 + 4 =	
42.	15 + 4 =	
43.	1 + 6 =	
44.	11 + 6 =	

EUREKA MATH®

Lesson 12:   Change 10 ones for 1 ten, 10 tens for 1 hundred, and 10 hundreds for 1 thousand.

© 2015 Great Minds®. eureka-math.org

A

Number Correct: _____

Place Value Counting to 100

1.	5 tens	
2.	6 tens 2 ones	
3.	6 tens 3 ones	
4.	6 tens 8 ones	
5.	60 + 4 =	
6.	4 + 60 =	
7.	8 tens	
8.	9 tens 4 ones	
9.	9 tens 5 ones	
10.	9 tens 8 ones	
11.	90 + 6 =	
12.	6 + 90 =	
13.	6 tens	
14.	7 tens 6 ones	
15.	7 tens 7 ones	
16.	7 tens 3 ones	
17.	70 + 8 =	
18.	8 + 70 =	
19.	9 tens	
20.	8 tens 1 one	
21.	8 tens 2 ones	
22.	8 tens 7 ones	

23.	80 + 4 =	
24.	4 + 80 =	
25.	7 tens	
26.	5 tens 8 ones	
27.	5 tens 9 ones	
28.	5 tens 2 ones	
29.	50 + 7 =	
30.	7 + 50 =	
31.	10 tens	
32.	7 tens 4 ones	
33.	80 + 3 =	
34.	7 + 90 =	
35.	6 tens + 10 =	
36.	9 tens 3 ones	
37.	70 + 2 =	
38.	3 + 50 =	
39.	60 + 2 tens =	
40.	8 tens 6 ones	
41.	90 + 2 =	
42.	5 + 60 =	
43.	8 tens 20 ones	
44.	30 + 7 tens =	

Lesson 13:   Read and write numbers within 1,000 after modeling with place value disks.

© 2015 Great Minds®. eureka-math.org

69

B

Number Correct: _____

Improvement: _____

Place Value Counting to 100

1.	6 tens	
2.	5 tens 2 ones	
3.	5 tens 3 ones	
4.	5 tens 8 ones	
5.	4 + 60 =	
6.	50 + 4 =	
7.	4 + 50 =	
8.	8 tens 4 ones	
9.	8 tens 5 ones	
10.	8 tens 8 ones	
11.	80 + 6 =	
12.	6 + 80 =	
13.	7 tens	
14.	9 tens 6 ones	
15.	9 tens 7 ones	
16.	9 tens 3 ones	
17.	90 + 8 =	
18.	8 + 90 =	
19.	5 tens	
20.	6 tens 1 one	
21.	6 tens 2 ones	
22.	6 tens 7 ones	

23.	60 + 4 =	
24.	4 + 60 =	
25.	8 tens	
26.	7 tens 8 ones	
27.	7 tens 9 ones	
28.	7 tens 2 ones	
29.	70 + 5 =	
30.	5 + 70 =	
31.	10 tens	
32.	5 tens 6 ones	
33.	60 + 3 =	
34.	6 + 70 =	
35.	5 tens + 10 =	
36.	7 tens 4 ones	
37.	80 + 3 =	
38.	2 + 90 =	
39.	70 + 2 tens	
40.	6 tens 8 ones	
41.	70 + 3 =	
42.	7 + 80 =	
43.	9 tens 10 ones	
44.	40 + 6 tens =	

EUREKA MATH®

Lesson 13: Read and write numbers within 1,000 after modeling with place value disks.

71

A

Number Correct: _____

Review of Subtraction in the Teens

1.	3 – 1 =	
2.	13 – 1 =	
3.	5 – 1 =	
4.	15 – 1 =	
5.	7 – 1 =	
6.	17 – 1 =	
7.	4 – 2 =	
8.	14 – 2 =	
9.	6 – 2 =	
10.	16 – 2 =	
11.	8 – 2 =	
12.	18 – 2 =	
13.	4 – 3 =	
14.	14 – 3 =	
15.	6 – 3 =	
16.	16 – 3 =	
17.	8 – 3 =	
18.	18 – 3 =	
19.	6 – 4 =	
20.	16 – 4 =	
21.	8 – 4 =	
22.	18 – 4 =	

23.	7 – 4 =	
24.	17 – 4 =	
25.	7 – 5 =	
26.	17 – 5 =	
27.	9 – 5 =	
28.	19 – 5 =	
29.	7 – 6 =	
30.	17 – 6 =	
31.	9 – 6 =	
32.	19 – 6 =	
33.	8 – 7 =	
34.	18 – 7 =	
35.	9 – 8 =	
36.	19 – 8 =	
37.	7 – 3 =	
38.	17 – 3 =	
39.	5 – 4 =	
40.	15 – 4 =	
41.	8 – 5 =	
42.	18 – 5 =	
43.	8 – 6 =	
44.	18 – 6 =	

EUREKA MATH®

Lesson 14:   Model numbers with more than 9 ones or 9 tens; write in expanded, unit, standard, and word forms.

73

© 2015 Great Minds®. eureka-math.org

B

Review of Subtraction in the Teens

Number Correct: _____

Improvement: _____

1.	2 – 1 =	
2.	12 – 1 =	
3.	4 – 1 =	
4.	14 – 1 =	
5.	6 – 1 =	
6.	16 – 1 =	
7.	3 – 2 =	
8.	13 – 2 =	
9.	5 – 2 =	
10.	15 – 2 =	
11.	7 – 2 =	
12.	17 – 2 =	
13.	5 – 3 =	
14.	15 – 3 =	
15.	7 – 3 =	
16.	17 – 3 =	
17.	9 – 3 =	
18.	19 – 3 =	
19.	5 – 4 =	
20.	15 – 4 =	
21.	7 – 4 =	
22.	17 – 4 =	

23.	9 – 4 =	
24.	19 – 4 =	
25.	6 – 5 =	
26.	16 – 5 =	
27.	8 – 5 =	
28.	18 – 5 =	
29.	8 – 6 =	
30.	18 – 6 =	
31.	9 – 6 =	
32.	19 – 6 =	
33.	9 – 7 =	
34.	19 – 7 =	
35.	9 – 8 =	
36.	19 – 8 =	
37.	8 – 3 =	
38.	18 – 3 =	
39.	6 – 4 =	
40.	16 – 4 =	
41.	9 – 5 =	
42.	19 – 5 =	
43.	7 – 6 =	
44.	17 – 6 =	

EUREKA MATH®

Lesson 14: Model numbers with more than 9 ones or 9 tens; write in expanded, unit, standard, and word forms.

75

A

Number Correct: _____

Expanded Notation

1.	20 + 1 =	
2.	20 + 2 =	
3.	20 + 3 =	
4.	20 + 9 =	
5.	30 + 9 =	
6.	40 + 9 =	
7.	80 + 9 =	
8.	40 + 4 =	
9.	50 + 5 =	
10.	10 + 7 =	
11.	20 + 5 =	
12.	200 + 30 =	
13.	300 + 40 =	
14.	400 + 50 =	
15.	500 + 60 =	
16.	600 + 70 =	
17.	700 + 80 =	
18.	200 + 30 + 5 =	
19.	300 + 40 + 5 =	
20.	400 + 50 + 6 =	
21.	500 + 60 + 7 =	
22.	600 + 70 + 8 =	

23.	400 + 20 + 5 =	
24.	200 + 60 + 1 =	
25.	200 + 1 =	
26.	300 + 1 =	
27.	400 + 1 =	
28.	500 + 1 =	
29.	700 + 1 =	
30.	300 + 50 + 2 =	
31.	300 + 2 =	
32.	100 + 10 + 7 =	
33.	100 + 7 =	
34.	700 + 10 + 5 =	
35.	700 + 5 =	
36.	300 + 40 + 7 =	
37.	300 + 7 =	
38.	500 + 30 + 2 =	
39.	500 + 2 =	
40.	2 + 500 =	
41.	2 + 600 =	
42.	2 + 40 + 600 =	
43.	3 + 10 + 700 =	
44.	8 + 30 + 700 =	

EUREKA MATH

Lesson 15: Explore a situation with more than 9 groups of ten.

77

B

Expanded Notation

Number Correct: _____

Improvement: _____

1.	10 + 1 =	
2.	10 + 2 =	
3.	10 + 3 =	
4.	10 + 9 =	
5.	20 + 9 =	
6.	30 + 9 =	
7.	70 + 9 =	
8.	30 + 3 =	
9.	40 + 4 =	
10.	80 + 7 =	
11.	90 + 5 =	
12.	100 + 20 =	
13.	200 + 30 =	
14.	300 + 40 =	
15.	400 + 50 =	
16.	500 + 60 =	
17.	600 + 70 =	
18.	300 + 40 + 5 =	
19.	400 + 50 + 6 =	
20.	500 + 60 + 7 =	
21.	600 + 70 + 8 =	
22.	700 + 80 + 9 =	

23.	500 + 30 + 6 =	
24.	300 + 70 + 1 =	
25.	300 + 1 =	
26.	400 + 1 =	
27.	500 + 1 =	
28.	600 + 1 =	
29.	900 + 1 =	
30.	400 + 60 + 3 =	
31.	400 + 3 =	
32.	100 + 10 + 5 =	
33.	100 + 5 =	
34.	800 + 10 + 5 =	
35.	800 + 5 =	
36.	200 + 30 + 7 =	
37.	200 + 7 =	
38.	600 + 40 + 2 =	
39.	600 + 2 =	
40.	2 + 600 =	
41.	3 + 600 =	
42.	3 + 40 + 600 =	
43.	5 + 10 + 800 =	
44.	9 + 20 + 700 =	

EUREKA MATH®

Lesson 15:   Explore a situation with more than 9 groups of ten.

79

A

Number Correct: _____

Sums—Crossing Ten

1.	9 + 1 =		23.	7 + 3 =	
2.	9 + 2 =		24.	7 + 4 =	
3.	9 + 3 =		25.	7 + 5 =	
4.	9 + 9 =		26.	7 + 9 =	
5.	8 + 2 =		27.	6 + 4 =	
6.	8 + 3 =		28.	6 + 5 =	
7.	8 + 4 =		29.	6 + 6 =	
8.	8 + 9 =		30.	6 + 9 =	
9.	9 + 1 =		31.	5 + 5 =	
10.	9 + 4 =		32.	5 + 6 =	
11.	9 + 5 =		33.	5 + 7 =	
12.	9 + 8 =		34.	5 + 9 =	
13.	8 + 2 =		35.	4 + 6 =	
14.	8 + 5 =		36.	4 + 7 =	
15.	8 + 6 =		37.	4 + 9 =	
16.	8 + 8 =		38.	3 + 7 =	
17.	9 + 1 =		39.	3 + 9 =	
18.	9 + 7 =		40.	5 + 8 =	
19.	8 + 2 =		41.	2 + 8 =	
20.	8 + 7 =		42.	4 + 8 =	
21.	9 + 1 =		43.	1 + 9 =	
22.	9 + 6 =		44.	2 + 9 =	

B

Number Correct: _____

Improvement: _____

Sums—Crossing Ten

1.	8 + 2 =	
2.	8 + 3 =	
3.	8 + 4 =	
4.	8 + 8 =	
5.	9 + 1 =	
6.	9 + 2 =	
7.	9 + 3 =	
8.	9 + 8 =	
9.	8 + 2 =	
10.	8 + 5 =	
11.	8 + 6 =	
12.	8 + 9 =	
13.	9 + 1 =	
14.	9 + 4 =	
15.	9 + 5 =	
16.	9 + 9 =	
17.	9 + 1 =	
18.	9 + 7 =	
19.	8 + 2 =	
20.	8 + 7 =	
21.	9 + 1 =	
22.	9 + 6 =	

23.	7 + 3 =	
24.	7 + 4 =	
25.	7 + 5 =	
26.	7 + 8 =	
27.	6 + 4 =	
28.	6 + 5 =	
29.	6 + 6 =	
30.	6 + 8 =	
31.	5 + 5 =	
32.	5 + 6 =	
33.	5 + 7 =	
34.	5 + 8 =	
35.	4 + 6 =	
36.	4 + 7 =	
37.	4 + 8 =	
38.	3 + 7 =	
39.	3 + 9 =	
40.	5 + 9 =	
41.	2 + 8 =	
42.	4 + 9 =	
43.	1 + 9 =	
44.	2 + 9 =	

A

Number Correct: _____

Sums—Crossing Ten

1.	9 + 2 =	
2.	9 + 3 =	
3.	9 + 4 =	
4.	9 + 7 =	
5.	7 + 9 =	
6.	10 + 1 =	
7.	10 + 2 =	
8.	10 + 3 =	
9.	10 + 8 =	
10.	8 + 10 =	
11.	8 + 3 =	
12.	8 + 4 =	
13.	8 + 5 =	
14.	8 + 9 =	
15.	9 + 8 =	
16.	7 + 4 =	
17.	10 + 5 =	
18.	6 + 5 =	
19.	7 + 5 =	
20.	9 + 5 =	
21.	5 + 9 =	
22.	10 + 6 =	

23.	4 + 7 =	
24.	4 + 8 =	
25.	5 + 6 =	
26.	5 + 7 =	
27.	3 + 8 =	
28.	3 + 9 =	
29.	2 + 9 =	
30.	5 + 10 =	
31.	5 + 8 =	
32.	9 + 6 =	
33.	6 + 9 =	
34.	7 + 6 =	
35.	6 + 7 =	
36.	8 + 6 =	
37.	6 + 8 =	
38.	8 + 7 =	
39.	7 + 8 =	
40.	6 + 6 =	
41.	7 + 7 =	
42.	8 + 8 =	
43.	9 + 9 =	
44.	4 + 9 =	

EUREKA MATH®

Lesson 17:   Compare two three-digit numbers using <, >, and = when there are more than 9 ones or 9 tens.

© 2015 Great Minds®. eureka-math.org

85

B

Number Correct: _____

Improvement: _____

Sums—Crossing Ten

1.	10 + 1 =		23.	5 + 6 =	
2.	10 + 2 =		24.	5 + 7 =	
3.	10 + 3 =		25.	4 + 7 =	
4.	10 + 9 =		26.	4 + 8 =	
5.	9 + 10 =		27.	4 + 10 =	
6.	9 + 2 =		28.	3 + 8 =	
7.	9 + 3 =		29.	3 + 9 =	
8.	9 + 4 =		30.	2 + 9 =	
9.	9 + 8 =		31.	5 + 8 =	
10.	8 + 9 =		32.	7 + 6 =	
11.	8 + 3 =		33.	6 + 7 =	
12.	8 + 4 =		34.	8 + 6 =	
13.	8 + 5 =		35.	6 + 8 =	
14.	8 + 7 =		36.	9 + 6 =	
15.	7 + 8 =		37.	6 + 9 =	
16.	7 + 4 =		38.	9 + 7 =	
17.	10 + 4 =		39.	7 + 9 =	
18.	6 + 5 =		40.	6 + 6 =	
19.	7 + 5 =		41.	7 + 7 =	
20.	9 + 5 =		42.	8 + 8 =	
21.	5 + 9 =		43.	9 + 9 =	
22.	10 + 8 =		44.	4 + 9 =	

EUREKA MATH®

Lesson 17: Compare two three-digit numbers using <, >, and = when there are more than 9 ones or 9 tens.

© 2015 Great Minds®. eureka-math.org

A

Number Correct: _____

Sums—Crossing Ten

1.	9 + 2 =	
2.	9 + 3 =	
3.	9 + 4 =	
4.	9 + 7 =	
5.	7 + 9 =	
6.	10 + 1 =	
7.	10 + 2 =	
8.	10 + 3 =	
9.	10 + 8 =	
10.	8 + 10 =	
11.	8 + 3 =	
12.	8 + 4 =	
13.	8 + 5 =	
14.	8 + 9 =	
15.	9 + 8 =	
16.	7 + 4 =	
17.	10 + 5 =	
18.	6 + 5 =	
19.	7 + 5 =	
20.	9 + 5 =	
21.	5 + 9 =	
22.	10 + 6 =	

23.	4 + 7 =	
24.	4 + 8 =	
25.	5 + 6 =	
26.	5 + 7 =	
27.	3 + 8 =	
28.	3 + 9 =	
29.	2 + 9 =	
30.	5 + 10 =	
31.	5 + 8 =	
32.	9 + 6 =	
33.	6 + 9 =	
34.	7 + 6 =	
35.	6 + 7 =	
36.	8 + 6 =	
37.	6 + 8 =	
38.	8 + 7 =	
39.	7 + 8 =	
40.	6 + 6 =	
41.	7 + 7 =	
42.	8 + 8 =	
43.	9 + 9 =	
44.	4 + 9 =	

EUREKA MATH

Lesson 18: Order numbers in different forms. (Optional)

89

© 2015 Great Minds®. eureka-math.org

B

Sums—Crossing Ten

1.	10 + 1 =		23.	5 + 6 =	
2.	10 + 2 =		24.	5 + 7 =	
3.	10 + 3 =		25.	4 + 7 =	
4.	10 + 9 =		26.	4 + 8 =	
5.	9 + 10 =		27.	4 + 10 =	
6.	9 + 2 =		28.	3 + 8 =	
7.	9 + 3 =		29.	3 + 9 =	
8.	9 + 4 =		30.	2 + 9 =	
9.	9 + 8 =		31.	5 + 8 =	
10.	8 + 9 =		32.	7 + 6 =	
11.	8 + 3 =		33.	6 + 7 =	
12.	8 + 4 =		34.	8 + 6 =	
13.	8 + 5 =		35.	6 + 8 =	
14.	8 + 7 =		36.	9 + 6 =	
15.	7 + 8 =		37.	6 + 9 =	
16.	7 + 4 =		38.	9 + 7 =	
17.	10 + 4 =		39.	7 + 9 =	
18.	6 + 5 =		40.	6 + 6 =	
19.	7 + 5 =		41.	7 + 7 =	
20.	9 + 5 =		42.	8 + 8 =	
21.	5 + 9 =		43.	9 + 9 =	
22.	10 + 8 =		44.	4 + 9 =	

EUREKA MATH®

Lesson 18: Order numbers in different forms. (Optional)

91

© 2015 Great Minds®. eureka-math.org

A

Number Correct: _____

Differences

1.	3 – 1 =	
2.	13 – 1 =	
3.	5 – 1 =	
4.	15 – 1 =	
5.	7 – 1 =	
6.	17 – 1 =	
7.	4 – 2 =	
8.	14 – 2 =	
9.	6 – 2 =	
10.	16 – 2 =	
11.	8 – 2 =	
12.	18 – 2 =	
13.	4 – 3 =	
14.	14 – 3 =	
15.	6 – 3 =	
16.	16 – 3 =	
17.	8 – 3 =	
18.	18 – 3 =	
19.	6 – 4 =	
20.	16 – 4 =	
21.	8 – 4 =	
22.	18 – 4 =	

23.	7 – 4 =	
24.	17 – 4 =	
25.	7 – 5 =	
26.	17 – 5 =	
27.	9 – 5 =	
28.	19 – 5 =	
29.	7 – 6 =	
30.	17 – 6 =	
31.	9 – 6 =	
32.	19 – 6 =	
33.	8 – 7 =	
34.	18 – 7 =	
35.	9 – 8 =	
36.	19 – 8 =	
37.	7 – 3 =	
38.	17 – 3 =	
39.	5 – 4 =	
40.	15 – 4 =	
41.	8 – 5 =	
42.	18 – 5 =	
43.	8 – 6 =	
44.	18 – 6 =	

EUREKA MATH®

Lesson 19: Model and use language to tell about 1 more and 1 less, 10 more and 10 less, and 100 more and 100 less.

© 2015 Great Minds®. eureka-math.org

93

B

Number Correct: _____

Differences

Improvement: _____

1.	2 – 1 =		23.	9 – 4 =	
2.	12 – 1 =		24.	19 – 4 =	
3.	4 – 1 =		25.	6 – 5 =	
4.	14 – 1 =		26.	16 – 5 =	
5.	6 – 1 =		27.	8 – 5 =	
6.	16 – 1 =		28.	18 – 5 =	
7.	3 – 2 =		29.	8 – 6 =	
8.	13 – 2 =		30.	18 – 6 =	
9.	5 – 2 =		31.	9 – 6 =	
10.	15 – 2 =		32.	19 – 6 =	
11.	7 – 2 =		33.	9 – 7 =	
12.	17 – 2 =		34.	19 – 7 =	
13.	5 – 3 =		35.	9 – 8 =	
14.	15 – 3 =		36.	19 – 8 =	
15.	7 – 3 =		37.	8 – 3 =	
16.	17 – 3 =		38.	18 – 3 =	
17.	9 – 3 =		39.	6 – 4 =	
18.	19 – 3 =		40.	16 – 4 =	
19.	5 – 4 =		41.	9 – 5 =	
20.	15 – 4 =		42.	19 – 5 =	
21.	7 – 4 =		43.	7 – 6 =	
22.	17 – 4 =		44.	17 – 6 =	

Lesson 19:   Model and use language to tell about 1 more and 1 less, 10 more and
10 less, and 100 more and 100 less.

95

A

Number Correct: _____

Differences

1.	3 – 1 =	
2.	13 – 1 =	
3.	5 – 1 =	
4.	15 – 1 =	
5.	7 – 1 =	
6.	17 – 1 =	
7.	4 – 2 =	
8.	14 – 2 =	
9.	6 – 2 =	
10.	16 – 2 =	
11.	8 – 2 =	
12.	18 – 2 =	
13.	4 – 3 =	
14.	14 – 3 =	
15.	6 – 3 =	
16.	16 – 3 =	
17.	8 – 3 =	
18.	18 – 3 =	
19.	6 – 4 =	
20.	16 – 4 =	
21.	8 – 4 =	
22.	18 – 4 =	

23.	7 – 4 =	
24.	17 – 4 =	
25.	7 – 5 =	
26.	17 – 5 =	
27.	9 – 5 =	
28.	19 – 5 =	
29.	7 – 6 =	
30.	17 – 6 =	
31.	9 – 6 =	
32.	19 – 6 =	
33.	8 – 7 =	
34.	18 – 7 =	
35.	9 – 8 =	
36.	19 – 8 =	
37.	7 – 3 =	
38.	17 – 3 =	
39.	5 – 4 =	
40.	15 – 4 =	
41.	8 – 5 =	
42.	18 – 5 =	
43.	8 – 6 =	
44.	18 – 6 =	

EUREKA MATH

Lesson 20:   Model 1 more and 1 less, 10 more and 10 less, and 100 more and 100 less when changing the hundreds place.

© 2015 Great Minds®. eureka-math.org

97

B

Number Correct: _____

Differences

Improvement: _____

1.	2 – 1 =		23.	9 – 4 =	
2.	12 – 1 =		24.	19 – 4 =	
3.	4 – 1 =		25.	6 – 5 =	
4.	14 – 1 =		26.	16 – 5 =	
5.	6 – 1 =		27.	8 – 5 =	
6.	16 – 1 =		28.	18 – 5 =	
7.	3 – 2 =		29.	8 – 6 =	
8.	13 – 2 =		30.	18 – 6 =	
9.	5 – 2 =		31.	9 – 6 =	
10.	15 – 2 =		32.	19 – 6 =	
11.	7 – 2 =		33.	9 – 7 =	
12.	17 – 2 =		34.	19 – 7 =	
13.	5 – 3 =		35.	9 – 8 =	
14.	15 – 3 =		36.	19 – 8 =	
15.	7 – 3 =		37.	8 – 3 =	
16.	17 – 3 =		38.	18 – 3 =	
17.	9 – 3 =		39.	6 – 4 =	
18.	19 – 3 =		40.	16 – 4 =	
19.	5 – 4 =		41.	9 – 5 =	
20.	15 – 4 =		42.	19 – 5 =	
21.	7 – 4 =		43.	7 – 6 =	
22.	17 – 4 =		44.	17 – 6 =	

EUREKA MATH

Lesson 20: Model 1 more and 1 less, 10 more and 10 less, and 100 more and 100 less when changing the hundreds place.

99

© 2015 Great Minds®. eureka-math.org

A

Number Correct: _____

Differences

1.	10 – 5 =		23.	11 – 3 =	
2.	10 – 0 =		24.	10 – 9 =	
3.	10 – 1 =		25.	11 – 9 =	
4.	10 – 9 =		26.	10 – 5 =	
5.	10 – 8 =		27.	11 – 5 =	
6.	10 – 2 =		28.	10 – 7 =	
7.	10 – 3 =		29.	11 – 7 =	
8.	10 – 7 =		30.	10 – 8 =	
9.	10 – 6 =		31.	11 – 8 =	
10.	10 – 4 =		32.	10 – 6 =	
11.	10 – 8 =		33.	11 – 6 =	
12.	10 – 3 =		34.	10 – 4 =	
13.	10 – 6 =		35.	11 – 4 =	
14.	10 – 9 =		36.	10 – 9 =	
15.	10 – 0 =		37.	12 – 9 =	
16.	10 – 5 =		38.	10 – 5 =	
17.	10 – 7 =		39.	12 – 5 =	
18.	10 – 2 =		40.	10 – 7 =	
19.	10 – 4 =		41.	12 – 7 =	
20.	10 – 1 =		42.	10 – 8 =	
21.	11 – 1 =		43.	12 – 8 =	
22.	11 – 2 =		44.	15 – 9 =	

B

Differences

Number Correct: _____

Improvement: _____

1.	10 – 0 =	
2.	10 – 5 =	
3.	10 – 9 =	
4.	10 – 1 =	
5.	10 – 2 =	
6.	10 – 8 =	
7.	10 – 7 =	
8.	10 – 3 =	
9.	10 – 4 =	
10.	10 – 6 =	
11.	10 – 2 =	
12.	10 – 7 =	
13.	10 – 4 =	
14.	10 – 1 =	
15.	10 – 0 =	
16.	10 – 5 =	
17.	10 – 3 =	
18.	10 – 8 =	
19.	10 – 6 =	
20.	10 – 9 =	
21.	11 – 1 =	
22.	11 – 2 =	

23.	11 – 3 =	
24.	10 – 5 =	
25.	11 – 5 =	
26.	10 – 9 =	
27.	11 – 9 =	
28.	10 – 8 =	
29.	11 – 8 =	
30.	10 – 7 =	
31.	11 – 7 =	
32.	10 – 4 =	
33.	11 – 4 =	
34.	10 – 6 =	
35.	11 – 6 =	
36.	10 – 5 =	
37.	12 – 5 =	
38.	10 – 9 =	
39.	12 – 9 =	
40.	10 – 8 =	
41.	12 – 8 =	
42.	10 – 7 =	
43.	12 – 7 =	
44.	14 – 9 =	

Grade 2
Module 4

A

Number Correct: _____

Add and Subtract Ones and Tens

1.	3 + 1 =		23.	50 + 30 =	
2.	30 + 10 =		24.	54 + 30 =	
3.	31 + 10 =		25.	54 + 3 =	
4.	31 + 1 =		26.	50 – 30 =	
5.	3 – 1 =		27.	59 – 30 =	
6.	30 – 10 =		28.	59 – 3 =	
7.	35 – 10 =		29.	67 + 30 =	
8.	35 – 1 =		30.	67 – 30 =	
9.	47 + 10 =		31.	67 – 3 =	
10.	10 – 1 =		32.	40 – 3 =	
11.	80 – 1 =		33.	42 – 3 =	
12.	40 + 20 =		34.	30 + 40 =	
13.	43 + 20 =		35.	32 + 40 =	
14.	43 + 2 =		36.	32 + 4 =	
15.	40 – 20 =		37.	70 – 40 =	
16.	45 – 20 =		38.	76 – 40 =	
17.	45 – 2 =		39.	76 – 4 =	
18.	57 + 2 =		40.	53 + 40 =	
19.	57 – 20 =		41.	53 + 4 =	
20.	10 – 2 =		42.	53 – 40 =	
21.	50 – 2 =		43.	90 – 4 =	
22.	51 – 2 =		44.	92 – 4 =	

Lesson 3:   Add and subtract multiples of 10 and some ones within 100.

B

Add and Subtract Ones and Tens

Number Correct: _____

Improvement: _____

1.	2 + 1 =		23.	40 + 30 =	
2.	20 + 10 =		24.	45 + 30 =	
3.	21 + 10 =		25.	45 + 3 =	
4.	21 + 1 =		26.	40 – 30 =	
5.	2 – 1 =		27.	49 – 30 =	
6.	20 – 10 =		28.	49 – 3 =	
7.	25 – 10 =		29.	57 + 30 =	
8.	25 – 1 =		30.	57 – 30 =	
9.	37 + 10 =		31.	57 – 3 =	
10.	10 – 1 =		32.	50 – 3 =	
11.	70 – 1 =		33.	52 – 3 =	
12.	50 + 20 =		34.	20 + 40 =	
13.	53 + 20 =		35.	23 + 40 =	
14.	53 + 2 =		36.	23 + 4 =	
15.	50 – 20 =		37.	80 – 40 =	
16.	54 – 20 =		38.	86 – 40 =	
17.	54 – 2 =		39.	86 – 4 =	
18.	64 + 2 =		40.	43 + 40 =	
19.	64 – 20 =		41.	43 + 4 =	
20.	10 – 2 =		42.	63 – 40 =	
21.	60 – 2 =		43.	80 – 4 =	
22.	61 – 2 =		44.	82 – 4 =	

EUREKA MATH®

Lesson 3:   Add and subtract multiples of 10 and some ones within 100.

109

© 2015 Great Minds®. eureka-math.org

A

Number Correct: _____

Sums to the Teens

1.	9 + 1 =	
2.	9 + 2 =	
3.	9 + 3 =	
4.	9 + 9 =	
5.	8 + 2 =	
6.	8 + 3 =	
7.	8 + 4 =	
8.	8 + 9 =	
9.	9 + 1 =	
10.	9 + 4 =	
11.	9 + 5 =	
12.	9 + 8 =	
13.	8 + 2 =	
14.	8 + 5 =	
15.	8 + 6 =	
16.	8 + 8 =	
17.	9 + 1 =	
18.	9 + 7 =	
19.	8 + 2 =	
20.	8 + 7 =	
21.	9 + 1 =	
22.	9 + 6 =	

23.	7 + 3 =	
24.	7 + 4 =	
25.	7 + 5 =	
26.	7 + 9 =	
27.	6 + 4 =	
28.	6 + 5 =	
29.	6 + 6 =	
30.	6 + 9 =	
31.	5 + 5 =	
32.	5 + 6 =	
33.	5 + 7 =	
34.	5 + 9 =	
35.	4 + 6 =	
36.	4 + 7 =	
37.	4 + 9 =	
38.	3 + 7 =	
39.	3 + 9 =	
40.	5 + 8 =	
41.	2 + 8 =	
42.	4 + 8 =	
43.	1 + 9 =	
44.	2 + 9 =	

EUREKA MATH®

Lesson 9:   Use math drawings to represent the composition when adding a
two-digit to a three digit addend.

111

© 2015 Great Minds®. eureka-math.org

B

Sums to the Teens

Number Correct: _____

Improvement: _____

1.	8 + 2 =	
2.	8 + 3 =	
3.	8 + 4 =	
4.	8 + 8 =	
5.	9 + 1 =	
6.	9 + 2 =	
7.	9 + 3 =	
8.	9 + 8 =	
9.	8 + 2 =	
10.	8 + 5 =	
11.	8 + 6 =	
12.	8 + 9 =	
13.	9 + 1 =	
14.	9 + 4 =	
15.	9 + 5 =	
16.	9 + 9 =	
17.	9 + 1 =	
18.	9 + 7 =	
19.	8 + 2 =	
20.	8 + 7 =	
21.	9 + 1 =	
22.	9 + 6 =	

23.	7 + 3 =	
24.	7 + 4 =	
25.	7 + 5 =	
26.	7 + 8 =	
27.	6 + 4 =	
28.	6 + 5 =	
29.	6 + 6 =	
30.	6 + 8 =	
31.	5 + 5 =	
32.	5 + 6 =	
33.	5 + 7 =	
34.	5 + 8 =	
35.	4 + 6 =	
36.	4 + 7 =	
37.	4 + 8 =	
38.	3 + 7 =	
39.	3 + 9 =	
40.	5 + 9 =	
41.	2 + 8 =	
42.	4 + 9 =	
43.	1 + 9 =	
44.	2 + 9 =	

Lesson 9:   Use math drawings to represent the composition when adding a two-digit to a three digit addend.

113

A

Number Correct: _____

Subtraction from Teens

1.	11 – 10 =	
2.	12 – 10 =	
3.	13 – 10 =	
4.	19 – 10 =	
5.	11 – 1 =	
6.	12 – 2 =	
7.	13 – 3 =	
8.	17 – 7 =	
9.	11 – 2 =	
10.	11 – 3 =	
11.	11 – 4 =	
12.	11 – 8 =	
13.	18 – 8 =	
14.	13 – 4 =	
15.	13 – 5 =	
16.	13 – 6 =	
17.	13 – 8 =	
18.	16 – 6 =	
19.	12 – 3 =	
20.	12 – 4 =	
21.	12 – 5 =	
22.	12 – 9 =	

23.	19 – 9 =	
24.	15 – 6 =	
25.	15 – 7 =	
26.	15 – 9 =	
27.	20 – 10 =	
28.	14 – 5 =	
29.	14 – 6 =	
30.	14 – 7 =	
31.	14 – 9 =	
32.	15 – 5 =	
33.	17 – 8 =	
34.	17 – 9 =	
35.	18 – 8 =	
36.	16 – 7 =	
37.	16 – 8 =	
38.	16 – 9 =	
39.	17 – 10 =	
40.	12 – 8 =	
41.	18 – 9 =	
42.	11 – 9 =	
43.	15 – 8 =	
44.	13 – 7 =	

EUREKA
MATH®

Lesson 10:   Use math drawings to represent the composition when adding a
two-digit to a three digit addend.

115

B

Subtraction from Teens

Number Correct: _____

Improvement: _____

1.	11 – 1 =	
2.	12 – 2 =	
3.	13 – 3 =	
4.	18 – 8 =	
5.	11 – 10 =	
6.	12 – 10 =	
7.	13 – 10 =	
8.	18 – 10 =	
9.	11 – 2 =	
10.	11 – 3 =	
11.	11 – 4 =	
12.	11 – 7 =	
13.	19 – 9 =	
14.	12 – 3 =	
15.	12 – 4 =	
16.	12 – 5 =	
17.	12 – 8 =	
18.	17 – 7 =	
19.	13 – 4 =	
20.	13 – 5 =	
21.	13 – 6 =	
22.	13 – 9 =	

23.	16 – 6 =	
24.	14 – 5 =	
25.	14 – 6 =	
26.	14 – 7 =	
27.	14 – 9 =	
28.	20 – 10 =	
29.	15 – 6 =	
30.	15 – 7 =	
31.	15 – 9 =	
32.	14 – 4 =	
33.	16 – 7 =	
34.	16 – 8 =	
35.	16 – 9 =	
36.	20 – 10 =	
37.	17 – 8 =	
38.	17 – 9 =	
39.	16 – 10 =	
40.	18 – 9 =	
41.	12 – 9 =	
42.	13 – 7 =	
43.	11 – 8 =	
44.	15 – 8 =	

EUREKA MATH®

Lesson 10:   Use math drawings to represent the composition when adding a two-digit to a three digit addend.

117

A

Number Correct: _____

Subtraction Patterns

1.	10 – 5 =	
2.	20 – 5 =	
3.	30 – 5 =	
4.	10 – 2 =	
5.	20 – 2 =	
6.	30 – 2 =	
7.	11 – 2 =	
8.	21 – 2 =	
9.	31 – 2 =	
10.	10 – 8 =	
11.	11 – 8 =	
12.	21 – 8 =	
13.	31 – 8 =	
14.	14 – 5 =	
15.	24 – 5 =	
16.	34 – 5 =	
17.	15 – 6 =	
18.	25 – 6 =	
19.	35 – 6 =	
20.	10 – 7 =	
21.	20 – 8 =	
22.	30 – 9 =	

23.	14 – 6 =	
24.	24 – 6 =	
25.	34 – 6 =	
26.	15 – 7 =	
27.	25 – 7 =	
28.	35 – 7 =	
29.	11 – 4 =	
30.	21 – 4 =	
31.	31 – 4 =	
32.	12 – 6 =	
33.	22 – 6 =	
34.	32 – 6 =	
35.	21 – 6 =	
36.	31 – 6 =	
37.	12 – 8 =	
38.	32 – 8 =	
39.	21 – 8 =	
40.	31 – 8 =	
41.	28 – 9 =	
42.	27 – 8 =	
43.	38 – 9 =	
44.	37 – 8 =	

EUREKA MATH

Lesson 13: Use math drawings to represent subtraction with and without
decomposition and relate drawings to a written method.

© 2015 Great Minds®. eureka-math.org

119

B

Subtraction Patterns

Number Correct: _____

Improvement: _____

1.	10 – 1 =	
2.	20 – 1 =	
3.	30 – 1 =	
4.	10 – 3 =	
5.	20 – 3 =	
6.	30 – 3 =	
7.	12 – 3 =	
8.	22 – 3 =	
9.	32 – 3 =	
10.	10 – 9 =	
11.	11 – 9 =	
12.	21 – 9 =	
13.	31 – 9 =	
14.	13 – 4 =	
15.	23 – 4 =	
16.	33 – 4 =	
17.	16 – 7 =	
18.	26 – 7 =	
19.	36 – 7 =	
20.	10 – 6 =	
21.	20 – 7 =	
22.	30 – 8 =	

23.	13 – 5 =	
24.	23 – 5 =	
25.	33 – 5 =	
26.	16 – 8 =	
27.	26 – 8 =	
28.	36 – 8 =	
29.	12 – 5 =	
30.	22 – 5 =	
31.	32 – 5 =	
32.	11 – 5 =	
33.	21 – 5 =	
34.	31 – 5 =	
35.	12 – 7 =	
36.	22 – 7 =	
37.	11 – 7 =	
38.	31 – 7 =	
39.	22 – 9 =	
40.	32 – 9 =	
41.	38 – 9 =	
42.	37 – 8 =	
43.	28 – 9 =	
44.	27 – 8 =	

Lesson 13: Use math drawings to represent subtraction with and without decomposition and relate drawings to a written method.

121

A

Number Correct: _____

Two-Digit Subtraction

1.	53 – 2 =	
2.	65 – 3 =	
3.	77 – 4 =	
4.	89 – 5 =	
5.	99 – 6 =	
6.	28 – 7 =	
7.	39 – 8 =	
8.	31 – 2 =	
9.	41 – 3 =	
10.	51 – 4 =	
11.	61 – 5 =	
12.	30 – 9 =	
13.	40 – 8 =	
14.	50 – 7 =	
15.	60 – 6 =	
16.	40 – 30 =	
17.	41 – 30 =	
18.	40 – 20 =	
19.	42 – 20 =	
20.	80 – 50 =	
21.	85 – 50 =	
22.	80 – 40 =	

23.	84 – 40 =	
24.	80 – 50 =	
25.	86 – 50 =	
26.	70 – 60 =	
27.	77 – 60 =	
28.	80 – 70 =	
29.	88 – 70 =	
30.	48 – 4 =	
31.	80 – 40 =	
32.	81 – 40 =	
33.	46 – 3 =	
34.	60 – 30 =	
35.	68 – 30 =	
36.	67 – 4 =	
37.	67 – 40 =	
38.	89 – 6 =	
39.	89 – 60 =	
40.	76 – 2 =	
41.	76 – 20 =	
42.	54 – 6 =	
43.	65 – 8 =	
44.	87 – 9 =	

EUREKA MATH®

Lesson 15: Represent subtraction with and without the decomposition when there is a three-digit minuend.

123

© 2015 Great Minds®. eureka-math.org

B

Number Correct: _____

Two-Digit Subtraction

Improvement: _____

1.	43 – 2 =		23.	94 – 50 =	
2.	55 – 3 =		24.	90 – 60 =	
3.	67 – 4 =		25.	96 – 60 =	
4.	79 – 5 =		26.	80 – 70 =	
5.	89 – 6 =		27.	87 – 70 =	
6.	98 – 7 =		28.	90 – 80 =	
7.	29 – 8 =		29.	98 – 80 =	
8.	21 – 2 =		30.	39 – 4 =	
9.	31 – 3 =		31.	90 – 40 =	
10.	41 – 4 =		32.	91 – 40 =	
11.	51 – 5 =		33.	47 – 3 =	
12.	20 – 9 =		34.	70 – 30 =	
13.	30 – 8 =		35.	78 – 30 =	
14.	40 – 7 =		36.	68 – 4 =	
15.	50 – 6 =		37.	68 – 40 =	
16.	30 – 20 =		38.	89 – 7 =	
17.	31 – 20 =		39.	89 – 70 =	
18.	50 – 30 =		40.	56 – 2 =	
19.	52 – 30 =		41.	56 – 20 =	
20.	70 – 40 =		42.	34 – 6 =	
21.	75 – 40 =		43.	45 – 8 =	
22.	90 – 50 =		44.	57 – 9 =	

EUREKA MATH

Lesson 15: Represent subtraction with and without the decomposition when there is a three-digit minuend.

125

A

Number Correct: _____

Addition Crossing a Ten

1.	38 + 1 =		23.	85 + 7 =	
2.	47 + 2 =		24.	85 + 9 =	
3.	56 + 3 =		25.	76 + 4 =	
4.	65 + 4 =		26.	76 + 5 =	
5.	31 + 8 =		27.	76 + 6 =	
6.	42 + 7 =		28.	76 + 9 =	
7.	53 + 6 =		29.	64 + 6 =	
8.	64 + 5 =		30.	64 + 7 =	
9.	49 + 1 =		31.	76 + 8 =	
10.	49 + 2 =		32.	43 + 7 =	
11.	49 + 3 =		33.	43 + 8 =	
12.	49 + 5 =		34.	43 + 9 =	
13.	58 + 2 =		35.	52 + 8 =	
14.	58 + 3 =		36.	52 + 9 =	
15.	58 + 4 =		37.	59 + 1 =	
16.	58 + 6 =		38.	59 + 3 =	
17.	67 + 3 =		39.	58 + 2 =	
18.	57 + 4 =		40.	58 + 4 =	
19.	57 + 5 =		41.	77 + 3 =	
20.	57 + 7 =		42.	77 + 5 =	
21.	85 + 5 =		43.	35 + 5 =	
22.	85 + 6 =		44.	35 + 8 =	

EUREKA MATH

B

Addition Crossing a Ten

Number Correct: _____

Improvement: _____

1.	28 + 1 =		23.	75 + 7 =	
2.	37 + 2 =		24.	75 + 9 =	
3.	46 + 3 =		25.	66 + 4 =	
4.	55 + 4 =		26.	66 + 5 =	
5.	21 + 8 =		27.	66 + 6 =	
6.	32 + 7 =		28.	66 + 9 =	
7.	43 + 6 =		29.	54 + 6 =	
8.	54 + 5 =		30.	54 + 7 =	
9.	39 + 1 =		31.	54 + 8 =	
10.	39 + 2 =		32.	33 + 7 =	
11.	39 + 3 =		33.	33 + 8 =	
12.	39 + 5 =		34.	33 + 9 =	
13.	48 + 2 =		35.	42 + 8 =	
14.	48 + 3 =		36.	42 + 9 =	
15.	48 + 4 =		37.	49 + 1 =	
16.	48 + 6 =		38.	49 + 3 =	
17.	57 + 3 =		39.	58 + 2 =	
18.	57 + 4 =		40.	58 + 4 =	
19.	57 + 5 =		41.	67 + 3 =	
20.	57 + 7 =		42.	67 + 5 =	
21.	75 + 5 =		43.	85 + 5 =	
22.	75 + 6 =		44.	85 + 8 =	

EUREKA MATH

Lesson 18: Use manipulatives to represent additions with two compositions.

129

© 2015 Great Minds®. eureka-math.org

A

Addition Crossing a Ten

Number Correct: _____

1.	38 + 1 =	
2.	47 + 2 =	
3.	56 + 3 =	
4.	65 + 4 =	
5.	31 + 8 =	
6.	42 + 7 =	
7.	53 + 6 =	
8.	64 + 5 =	
9.	49 + 1 =	
10.	49 + 2 =	
11.	49 + 3 =	
12.	49 + 5 =	
13.	58 + 2 =	
14.	58 + 3 =	
15.	58 + 4 =	
16.	58 + 6 =	
17.	67 + 3 =	
18.	57 + 4 =	
19.	57 + 5 =	
20.	57 + 7 =	
21.	85 + 5 =	
22.	85 + 6 =	

23.	85 + 7 =	
24.	85 + 9 =	
25.	76 + 4 =	
26.	76 + 5 =	
27.	76 + 6 =	
28.	76 + 9 =	
29.	64 + 6 =	
30.	64 + 7 =	
31.	76 + 8 =	
32.	43 + 7 =	
33.	43 + 8 =	
34.	43 + 9 =	
35.	52 + 8 =	
36.	52 + 9 =	
37.	59 + 1 =	
38.	59 + 3 =	
39.	58 + 2 =	
40.	58 + 4 =	
41.	77 + 3 =	
42.	77 + 5 =	
43.	35 + 5 =	
44.	35 + 8 =	

EUREKA MATH®

Lesson 20: Use math drawings to represent additions with up to two compositions and relate drawings to a written method.

© 2015 Great Minds®. eureka-math.org

131

B

Number Correct: _____

Improvement: _____

Addition Crossing a Ten

1.	28 + 1 =		23.	75 + 7 =	
2.	37 + 2 =		24.	75 + 9 =	
3.	46 + 3 =		25.	66 + 4 =	
4.	55 + 4 =		26.	66 + 5 =	
5.	21 + 8 =		27.	66 + 6 =	
6.	32 + 7 =		28.	66 + 9 =	
7.	43 + 6 =		29.	54 + 6 =	
8.	54 + 5 =		30.	54 + 7 =	
9.	39 + 1 =		31.	54 + 8 =	
10.	39 + 2 =		32.	33 + 7 =	
11.	39 + 3 =		33.	33 + 8 =	
12.	39 + 5 =		34.	33 + 9 =	
13.	48 + 2 =		35.	42 + 8 =	
14.	48 + 3 =		36.	42 + 9 =	
15.	48 + 4 =		37.	49 + 1 =	
16.	48 + 6 =		38.	49 + 3 =	
17.	57 + 3 =		39.	58 + 2 =	
18.	57 + 4 =		40.	58 + 4 =	
19.	57 + 5 =		41.	67 + 3 =	
20.	57 + 7 =		42.	67 + 5 =	
21.	75 + 5 =		43.	85 + 5 =	
22.	75 + 6 =		44.	85 + 8 =	

EUREKA MATH

Lesson 20: Use math drawings to represent additions with up to two compositions and relate drawings to a written method.

133

© 2015 Great Minds®. eureka-math.org

A

Number Correct: _____

Subtraction Patterns

1.	10 – 1 =		23.	21 – 6 =	
2.	10 – 2 =		24.	91 – 6 =	
3.	20 – 2 =		25.	10 – 7 =	
4.	40 – 2 =		26.	11 – 7 =	
5.	10 – 2 =		27.	31 – 7 =	
6.	11 – 2 =		28.	10 – 8 =	
7.	21 – 2 =		29.	11 – 8 =	
8.	51 – 2 =		30.	41 – 8 =	
9.	10 – 3 =		31.	10 – 9 =	
10.	11 – 3 =		32.	11 – 9 =	
11.	21 – 3 =		33.	51 – 9 =	
12.	61 – 3 =		34.	12 – 3 =	
13.	10 – 4 =		35.	82 – 3 =	
14.	11 – 4 =		36.	13 – 5 =	
15.	21 – 4 =		37.	73 – 5 =	
16.	71 – 4 =		38.	14 – 6 =	
17.	10 – 5 =		39.	84 – 6 =	
18.	11 – 5 =		40.	15 – 8 =	
19.	21 – 5 =		41.	95 – 8 =	
20.	81 – 5 =		42.	16 – 7 =	
21.	10 – 6 =		43.	46 – 7 =	
22.	11 – 6 =		44.	68 – 9 =	

EUREKA MATH

Lesson 23: Use number bonds to break apart three-digit minuends and subtract from the hundred.

135

© 2015 Great Minds®. eureka-math.org

B

Subtraction Patterns

Number Correct: _____

Improvement: _____

1.	10 – 2 =	
2.	20 – 2 =	
3.	30 – 2 =	
4.	50 – 2 =	
5.	10 – 2 =	
6.	11 – 2 =	
7.	21 – 2 =	
8.	61 – 2 =	
9.	10 – 3 =	
10.	11 – 3 =	
11.	21 – 3 =	
12.	71 – 3 =	
13.	10 – 4 =	
14.	11 – 4 =	
15.	21 – 4 =	
16.	81 – 4 =	
17.	10 – 5 =	
18.	11 – 5 =	
19.	21 – 5 =	
20.	91 – 5 =	
21.	10 – 6 =	
22.	11 – 6 =	

23.	21 – 6 =	
24.	41 – 6 =	
25.	10 – 7 =	
26.	11 – 7 =	
27.	51 – 7 =	
28.	10 – 8 =	
29.	11 – 8 =	
30.	61 – 8 =	
31.	10 – 9 =	
32.	11 – 9 =	
33.	31 – 9 =	
34.	12 – 3 =	
35.	92 – 3 =	
36.	13 – 5 =	
37.	43 – 5 =	
38.	14 – 6 =	
39.	64 – 6 =	
40.	15 – 8 =	
41.	85 – 8 =	
42.	16 – 7 =	
43.	76 – 7 =	
44.	58 – 9 =	

EUREKA MATH®

Lesson 23: Use number bonds to break apart three-digit minuends and subtract from the hundred.

A

Number Correct: _____

Subtraction Patterns

1.	30 – 1 =	
2.	40 – 2 =	
3.	50 – 3 =	
4.	50 – 4 =	
5.	50 – 5 =	
6.	50 – 9 =	
7.	51 – 9 =	
8.	61 – 9 =	
9.	81 – 9 =	
10.	82 – 9 =	
11.	92 – 9 =	
12.	93 – 9 =	
13.	93 – 8 =	
14.	83 – 8 =	
15.	33 – 8 =	
16.	33 – 7 =	
17.	43 – 7 =	
18.	53 – 6 =	
19.	63 – 6 =	
20.	63 – 5 =	
21.	73 – 5 =	
22.	93 – 5 =	

23.	31 – 2 =	
24.	31 – 3 =	
25.	31 – 4 =	
26.	41 – 4 =	
27.	51 – 5 =	
28.	61 – 6 =	
29.	71 – 7 =	
30.	81 – 8 =	
31.	82 – 8=	
32.	82 – 7 =	
33.	82 – 6 =	
34.	82 – 3 =	
35.	34 – 5 =	
36.	45 – 6 =	
37.	56 – 7 =	
38.	67 – 8 =	
39.	78 – 9 =	
40.	77 – 9 =	
41.	64 – 6 =	
42.	24 – 8 =	
43.	35 – 8 =	
44.	36 – 8 =	

Lesson 26:   Use math drawings to represent subtraction with up to two
decompositions and relate drawings to a written method.

EUREKA MATH

139

© 2015 Great Minds®. eureka-math.org

B

Subtraction Patterns

Number Correct: _____

Improvement: _____

1.	20 – 1 =	
2.	30 – 2 =	
3.	40 – 3 =	
4.	40 – 4 =	
5.	40 – 5 =	
6.	40 – 9 =	
7.	41 – 9 =	
8.	51 – 9 =	
9.	71 – 9 =	
10.	72	
11.	82	
12.	83	
13.	83 – 8 =	
14.	93 – 8 =	
15.	23 – 8 =	
16.	23 – 7 =	
17.	33 – 7 =	
18.	43 – 6 =	
19.	53 – 6 =	
20.	53 – 5 =	
21.	63 – 5 =	
22.	83 – 5 =	

23.	21 – 2 =	
24.	21 – 3 =	
25.	21 – 4 =	
26.	31 – 4 =	
27.	41 – 5 =	
28.	51 – 6 =	
29.	61 – 7 =	
30.	71 – 8 =	
31.	72 – 8 =	
32.	72 – 7 =	
33.	72 – 6 =	
34.	72 – 3 =	
35.	24 – 5 =	
36.	35 – 6 =	
37.	46 – 7 =	
38.	57 – 8 =	
39.	68 – 9 =	
40.	67 – 9 =	
41.	54 – 6 =	
42.	24 – 9 =	
43.	35 – 9 =	
44.	46 – 9 =	

EUREKA MATH

Lesson 26: Use math drawings to represent subtraction with up to two decompositions and relate drawings to a written method.

141

© 2015 Great Minds®. eureka-math.org

A

Number Correct: _____

Subtraction from a Ten or a Hundred

1.	10 – 1 =	
2.	100 – 10 =	
3.	90 – 1 =	
4.	100 – 11 =	
5.	10 – 2 =	
6.	100 – 20 =	
7.	80 – 1 =	
8.	100 – 21 =	
9.	10 – 5 =	
10.	100 – 50 =	
11.	50 – 2 =	
12.	100 – 52 =	
13.	10 – 4 =	
14.	100 – 40 =	
15.	60 – 1 =	
16.	100 – 41 =	
17.	10 – 3 =	
18.	100 – 30 =	
19.	70 – 5 =	
20.	100 – 35 =	
21.	100 – 80 =	
22.	100 – 81 =	

23.	100 – 82 =	
24.	100 – 85 =	
25.	100 – 15 =	
26.	100 – 70 =	
27.	100 – 71 =	
28.	100 – 72 =	
29.	100 – 75 =	
30.	100 – 25 =	
31.	100 – 10 =	
32.	100 – 11 =	
33.	100 – 12 =	
34.	100 – 18 =	
35.	100 – 82 =	
36.	100 – 60 =	
37.	100 – 6 =	
38.	100 – 70 =	
39.	100 – 7 =	
40.	100 – 43 =	
41.	100 – 8 =	
42.	100 – 59 =	
43.	100 – 4 =	
44.	100 – 68 =	

Lesson 27: Subtract from 200 and from numbers with zeros in the tens place.

143

EUREKA MATH

B

Subtraction from a Ten or a Hundred

Number Correct: _____

Improvement: _____

1.	10 – 5 =	
2.	100 – 50 =	
3.	50 – 1 =	
4.	100 – 51 =	
5.	10 – 2 =	
6.	100 – 20 =	
7.	80 – 1 =	
8.	100 – 21 =	
9.	10 – 1 =	
10.	100 – 10 =	
11.	90 – 2 =	
12.	100 – 12 =	
13.	10 – 3 =	
14.	100 – 30 =	
15.	70 – 1 =	
16.	100 – 31 =	
17.	10 – 4 =	
18.	100 – 40 =	
19.	60 – 5 =	
20.	100 – 45 =	
21.	100 – 70 =	
22.	100 – 71 =	

23.	100 – 72 =	
24.	100 – 75 =	
25.	100 – 25 =	
26.	100 – 80 =	
27.	100 – 81 =	
28.	100 – 82 =	
29.	100 – 85 =	
30.	100 – 15 =	
31.	100 – 10 =	
32.	100 – 11 =	
33.	100 – 12 =	
34.	100 – 17 =	
35.	100 – 83 =	
36.	100 – 70 =	
37.	100 – 7 =	
38.	100 – 60 =	
39.	100 – 6 =	
40.	100 – 42 =	
41.	100 – 4 =	
42.	100 – 58 =	
43.	100 – 8 =	
44.	100 – 67 =	

Lesson 27:   Subtract from 200 and from numbers with zeros in the tens place.   145

A

Number Correct: _____

Subtraction Crossing a Ten

1.	30 – 1 =	
2.	40 – 2 =	
3.	50 – 3 =	
4.	50 – 4 =	
5.	50 – 5 =	
6.	50 – 9 =	
7.	51 – 9 =	
8.	61 – 9 =	
9.	81 – 9 =	
10.	82 – 9 =	
11.	92 – 9 =	
12.	93 – 9 =	
13.	93 – 8 =	
14.	83 – 8 =	
15.	33 – 8 =	
16.	33 – 7 =	
17.	43 – 7 =	
18.	53 – 6 =	
19.	63 – 6 =	
20.	63 – 5 =	
21.	73 – 5 =	
22.	93 – 5 =	

23.	31 – 2 =	
24.	31 – 3 =	
25.	31 – 4 =	
26.	41 – 4 =	
27.	51 – 5 =	
28.	61 – 6 =	
29.	71 – 7 =	
30.	81 – 8 =	
31.	82 – 8 =	
32.	82 – 7 =	
33.	82 – 6 =	
34.	82 – 3 =	
35.	34 – 5 =	
36.	45 – 6 =	
37.	56 – 7 =	
38.	67 – 8 =	
39.	78 – 9 =	
40.	77 – 9 =	
41.	64 – 6 =	
42.	24 – 8 =	
43.	35 – 8 =	
44.	36 – 8 =	

Lesson 30:   Compare totals below to new groups below as written methods.

147

B

Number Correct: _____

Improvement: _____

Subtraction Crossing a Ten

1.	20 – 1 =	
2.	30 – 2 =	
3.	40 – 3 =	
4.	40 – 4 =	
5.	40 – 5 =	
6.	40 – 9 =	
7.	41 – 9 =	
8.	51 – 9 =	
9.	71 – 9 =	
10.	72 – 9 =	
11.	82 – 9 =	
12.	83 – 9 =	
13.	83 – 8 =	
14.	93 – 8 =	
15.	23 – 8 =	
16.	23 – 7 =	
17.	33 – 7 =	
18.	43 – 6 =	
19.	53 – 6 =	
20.	53 – 5 =	
21.	63 – 5 =	
22.	83 – 5 =	

23.	21 – 2 =	
24.	21 – 3 =	
25.	21 – 4 =	
26.	31 – 4 =	
27.	41 – 5 =	
28.	51 – 6 =	
29.	61 – 7 =	
30.	71 – 8 =	
31.	72 – 8 =	
32.	72 – 7 =	
33.	72 – 6 =	
34.	72 – 3 =	
35.	24 – 5 =	
36.	35 – 6 =	
37.	46 – 7 =	
38.	57 – 8 =	
39.	68 – 9 =	
40.	67 – 9 =	
41.	54 – 6 =	
42.	24 – 9 =	
43.	35 – 9 =	
44.	46 – 9 =	

EUREKA MATH

Lesson 30:   Compare totals below to new groups below as written methods.

149

© 2015 Great Minds®. eureka-math.org

Grade 2
Module 5

A

Number Correct: _____

Adding Multiples of Ten and Some Ones

1.	40 + 3 =		23.	45 + 44 =	
2.	40 + 8 =		24.	44 + 45 =	
3.	40 + 9 =		25.	30 + 20 =	
4.	40 + 10 =		26.	34 + 20 =	
5.	41 + 10 =		27.	34 + 21 =	
6.	42 + 10 =		28.	34 + 25 =	
7.	45 + 10 =		29.	34 + 52 =	
8.	45 + 11 =		30.	50 + 30 =	
9.	45 + 12 =		31.	56 + 30 =	
10.	44 + 12 =		32.	56 + 31 =	
11.	43 + 12 =		33.	56 + 32 =	
12.	43 + 13 =		34.	32 + 56 =	
13.	13 + 43 =		35.	23 + 56 =	
14.	40 + 20 =		36.	24 + 75 =	
15.	41 + 20 =		37.	16 + 73 =	
16.	42 + 20 =		38.	34 + 54 =	
17.	47 + 20 =		39.	62 + 37 =	
18.	47 + 30 =		40.	45 + 34 =	
19.	47 + 40 =		41.	27 + 61 =	
20.	47 + 41 =		42.	16 + 72 =	
21.	47 + 42 =		43.	36 + 42 =	
22.	45 + 42 =		44.	32 + 54 =	

B

Number Correct: _____

Improvement: _____

Adding Multiples of Ten and Some Ones

1.	50 + 3 =	
2.	50 + 8 =	
3.	50 + 9 =	
4.	50 + 10 =	
5.	51 + 10 =	
6.	52 + 10 =	
7.	55 + 10 =	
8.	55 + 11 =	
9.	55 + 12 =	
10.	54 + 12 =	
11.	53 + 12 =	
12.	53 + 13 =	
13.	13 + 43 =	
14.	50 + 20 =	
15.	51 + 20 =	
16.	52 + 20 =	
17.	57 + 20 =	
18.	57 + 30 =	
19.	57 + 40 =	
20.	57 + 41 =	
21.	57 + 42 =	
22.	55 + 42 =	

23.	55 + 44 =	
24.	44 + 55 =	
25.	40 + 20 =	
26.	44 + 20 =	
27.	44 + 21 =	
28.	44 + 25 =	
29.	44 + 52 =	
30.	60 + 30 =	
31.	66 + 30 =	
32.	66 + 31 =	
33.	66 + 32 =	
34.	32 + 66 =	
35.	23 + 66 =	
36.	25 + 74 =	
37.	13 + 76 =	
38.	43 + 45 =	
39.	26 + 73 =	
40.	54 + 43 =	
41.	72 + 16 =	
42.	61 + 27 =	
43.	63 + 24 =	
44.	32 + 45 =	

A

Number Correct: _____

Subtracting Multiples of Ten and Some Ones

1.	33 – 22 =		23.	99 – 32 =	
2.	44 – 33 =		24.	86 – 32 =	
3.	55 – 44 =		25.	79 – 32 =	
4.	99 – 88 =		26.	79 – 23 =	
5.	33 – 11 =		27.	68 – 13 =	
6.	44 – 22 =		28.	69 – 23 =	
7.	55 – 33 =		29.	89 – 14 =	
8.	88 – 22 =		30.	77 – 12 =	
9.	66 – 22 =		31.	57 – 12 =	
10.	43 – 11 =		32.	77 – 32 =	
11.	34 – 11 =		33.	99 – 36 =	
12.	45 – 11 =		34.	88 – 25 =	
13.	46 – 12 =		35.	89 – 36 =	
14.	55 – 12 =		36.	98 – 16 =	
15.	54 – 12 =		37.	78 – 26 =	
16.	55 – 21 =		38.	99 – 37 =	
17.	64 – 21 =		39.	89 – 38 =	
18.	63 – 21 =		40.	59 – 28 =	
19.	45 – 21 =		41.	99 – 58 =	
20.	34 – 12 =		42.	99 – 45 =	
21.	43 – 21 =		43.	78 – 43 =	
22.	54 – 32 =		44.	98 – 73 =	

EUREKA MATH®

B

Number Correct: _____

Improvement: _____

Subtracting Multiples of Ten and Some Ones

1.	33 – 11 =		23.	99 – 42 =	
2.	44 – 11 =		24.	79 – 32 =	
3.	55 – 11 =		25.	89 – 52 =	
4.	88 – 11 =		26.	99 – 23 =	
5.	33 – 22 =		27.	79 – 13 =	
6.	44 – 22 =		28.	79 – 23 =	
7.	55 – 22 =		29.	99 – 14 =	
8.	99 – 22 =		30.	87 – 12 =	
9.	77 – 22 =		31.	77 – 12 =	
10.	34 – 11 =		32.	87 – 32 =	
11.	43 – 11 =		33.	99 – 36 =	
12.	54 – 11 =		34.	78 – 25 =	
13.	55 – 12 =		35.	79 – 36 =	
14.	46 – 12 =		36.	88 – 16 =	
15.	44 – 12 =		37.	88 – 26 =	
16.	64 – 21 =		38.	89 – 37 =	
17.	55 – 21 =		39.	99 – 38 =	
18.	53 – 21 =		40.	69 – 28 =	
19.	44 – 21 =		41.	89 – 58 =	
20.	34 – 22 =		42.	99 – 45 =	
21.	43 – 22 =		43.	68 – 43 =	
22.	54 – 22 =		44.	98 – 72 =	

EUREKA MATH®

Lesson 4:   Subtract multiples of 100 and some tens within 1,000.

159

© 2015 Great Minds®. eureka-math.org

A

Number Correct: _____

Two-Digit Addition

1.	38 + 1 =		23.	85 + 7 =	
2.	47 + 2 =		24.	85 + 9 =	
3.	56 + 3 =		25.	76 + 4 =	
4.	65 + 4 =		26.	76 + 5 =	
5.	31 + 8 =		27.	76 + 6 =	
6.	42 + 7 =		28.	76 + 9 =	
7.	53 + 6 =		29.	64 + 6 =	
8.	64 + 5 =		30.	64 + 7 =	
9.	49 + 1 =		31.	76 + 8 =	
10.	49 + 2 =		32.	43 + 7 =	
11.	49 + 3 =		33.	43 + 8 =	
12.	49 + 5 =		34.	43 + 9 =	
13.	58 + 2 =		35.	52 + 8 =	
14.	58 + 3 =		36.	52 + 9 =	
15.	58 + 4 =		37.	59 + 1 =	
16.	58 + 6 =		38.	59 + 3 =	
17.	67 + 3 =		39.	58 + 2 =	
18.	57 + 4 =		40.	58 + 4 =	
19.	57 + 5 =		41.	77 + 3 =	
20.	57 + 7 =		42.	77 + 5 =	
21.	85 + 5 =		43.	35 + 5 =	
22.	85 + 6 =		44.	35 + 8 =	

B

Number Correct: _____

Improvement: _____

Two-Digit Addition

1.	28 + 1 =	
2.	37 + 2 =	
3.	46 + 3 =	
4.	55 + 4 =	
5.	21 + 8 =	
6.	32 + 7 =	
7.	43 + 6 =	
8.	54 + 5 =	
9.	39 + 1 =	
10.	39 + 2 =	
11.	39 + 3 =	
12.	39 + 5 =	
13.	48 + 2 =	
14.	48 + 3 =	
15.	48 + 4 =	
16.	48 + 6 =	
17.	57 + 3 =	
18.	57 + 4 =	
19.	57 + 5 =	
20.	57 + 7 =	
21.	75 + 5 =	
22.	75 + 6 =	

23.	75 + 7 =	
24.	75 + 9 =	
25.	66 + 4 =	
26.	66 + 5 =	
27.	66 + 6 =	
28.	66 + 9 =	
29.	54 + 6 =	
30.	54 + 7 =	
31.	54 + 8 =	
32.	33 + 7 =	
33.	33 + 8 =	
34.	33 + 9 =	
35.	42 + 8 =	
36.	42 + 9 =	
37.	49 + 1 =	
38.	49 + 3 =	
39.	58 + 2 =	
40.	58 + 4 =	
41.	67 + 3 =	
42.	67 + 5 =	
43.	85 + 5 =	
44.	85 + 8 =	

A

Number Correct: _____

Addition Crossing Tens

1.	8 + 2 =		23.	18 + 6 =	
2.	18 + 2 =		24.	28 + 6 =	
3.	38 + 2 =		25.	16 + 8 =	
4.	7 + 3 =		26.	26 + 8 =	
5.	17 + 3 =		27.	18 + 7 =	
6.	37 + 3 =		28.	18 + 8 =	
7.	8 + 3 =		29.	28 + 7 =	
8.	18 + 3 =		30.	28 + 8 =	
9.	28 + 3 =		31.	15 + 9 =	
10.	6 + 5 =		32.	16 + 9 =	
11.	16 + 5 =		33.	25 + 9 =	
12.	26 + 5 =		34.	26 + 9 =	
13.	18 + 4 =		35.	14 + 7 =	
14.	28 + 4 =		36.	16 + 6 =	
15.	16 + 6 =		37.	15 + 8 =	
16.	26 + 6 =		38.	23 + 8 =	
17.	18 + 5 =		39.	25 + 7 =	
18.	28 + 5 =		40.	15 + 7 =	
19.	16 + 7 =		41.	24 + 7 =	
20.	26 + 7 =		42.	14 + 9 =	
21.	19 + 2 =		43.	19 + 8 =	
22.	17 + 4 =		44.	28 + 9 =	

EUREKA MATH®

Lesson 10:   Use math drawings to represent additions with up to two
compositions and relate drawings to the addition algorithm.

165

© 2015 Great Minds®. eureka-math.org

B

Number Correct: _____

Improvement: _____

Addition Crossing Tens

1.	9 + 1 =		23.	19 + 5 =	
2.	19 + 1 =		24.	29 + 5 =	
3.	39 + 1 =		25.	17 + 7 =	
4.	6 + 4 =		26.	27 + 7 =	
5.	16 + 4 =		27.	19 + 6 =	
6.	36 + 4 =		28.	19 + 7 =	
7.	9 + 2 =		29.	29 + 6 =	
8.	19 + 2 =		30.	29 + 7 =	
9.	29 + 2 =		31.	17 + 8 =	
10.	7 + 4 =		32.	17 + 9 =	
11.	17 + 4 =		33.	27 + 8 =	
12.	27 + 4 =		34.	27 + 9 =	
13.	19 + 3 =		35.	12 + 9 =	
14.	29 + 3 =		36.	14 + 8 =	
15.	17 + 5 =		37.	16 + 7 =	
16.	27 + 5 =		38.	28 + 6 =	
17.	19 + 4 =		39.	26 + 8 =	
18.	29 + 4 =		40.	24 + 8 =	
19.	17 + 6 =		41.	13 + 8 =	
20.	27 + 6 =		42.	24 + 9 =	
21.	18 + 3 =		43.	29 + 8 =	
22.	26 + 5 =		44.	18 + 9 =	

EUREKA MATH®

Lesson 10: Use math drawings to represent additions with up to two compositions and relate drawings to the addition algorithm.

167

© 2015 Great Minds®. eureka-math.org

A

Number Correct: _____

Compensation Addition

1.	98 + 3 =		23.	99 + 12 =	
2.	98 + 4 =		24.	99 + 23 =	
3.	98 + 5 =		25.	99 + 34 =	
4.	98 + 8 =		26.	99 + 45 =	
5.	98 + 6 =		27.	99 + 56 =	
6.	98 + 9 =		28.	99 + 67 =	
7.	98 + 7 =		29.	99 + 78 =	
8.	99 + 2 =		30.	35 + 99 =	
9.	99 + 3 =		31.	45 + 98 =	
10.	99 + 4 =		32.	46 + 99 =	
11.	99 + 9 =		33.	56 + 98 =	
12.	99 + 6 =		34.	67 + 99 =	
13.	99 + 8 =		35.	77 + 98 =	
14.	99 + 5 =		36.	68 + 99 =	
15.	99 + 7 =		37.	78 + 98 =	
16.	98 + 13 =		38.	99 + 95 =	
17.	98 + 24 =		39.	93 + 99 =	
18.	98 + 35 =		40.	99 + 95 =	
19.	98 + 46 =		41.	94 + 99 =	
20.	98 + 57 =		42.	98 + 96 =	
21.	98 + 68 =		43.	94 + 98 =	
22.	98 + 79 =		44.	98 + 88 =	

B

Number Correct: _____

Improvement: _____

Compensation Addition

1.	99 + 2 =	
2.	99 + 3 =	
3.	99 + 4 =	
4.	99 + 8 =	
5.	99 + 6 =	
6.	99 + 9 =	
7.	99 + 5 =	
8.	99 + 7 =	
9.	98 + 3 =	
10.	98 + 4 =	
11.	98 + 5 =	
12.	98 + 9 =	
13.	98 + 7 =	
14.	98 + 8 =	
15.	98 + 6 =	
16.	99 + 12 =	
17.	99 + 23 =	
18.	99 + 34 =	
19.	99 + 45 =	
20.	99 + 56 =	
21.	99 + 67 =	
22.	99 + 78 =	

23.	98 + 13 =	
24.	98 + 24 =	
25.	98 + 35 =	
26.	98 + 46 =	
27.	98 + 57 =	
28.	98 + 68 =	
29.	98 + 79 =	
30.	25 + 99 =	
31.	35 + 98 =	
32.	36 + 99 =	
33.	46 + 98 =	
34.	57 + 99 =	
35.	67 + 98 =	
36.	78 + 99 =	
37.	88 + 98 =	
38.	99 + 93 =	
39.	95 + 99 =	
40.	99 + 97 =	
41.	92 + 99 =	
42.	98 + 94 =	
43.	96 + 98 =	
44.	98 + 86 =	

EUREKA MATH®

Lesson 12: Choose and explain solution strategies and record with a written addition method.

171

© 2015 Great Minds®. eureka-math.org

Name _____   Date _____

1.	10 + 2 =	21.	2 + 9 =
2.	10 + 5 =	22.	4 + 8 =
3.	10 + 1 =	23.	5 + 9 =
4.	8 + 10 =	24.	6 + 6 =
5.	7 + 10 =	25.	7 + 5 =
6.	10 + 3 =	26.	5 + 8 =
7.	12 + 2 =	27.	8 + 3 =
8.	14 + 3 =	28.	6 + 8 =
9.	15 + 4 =	29.	4 + 6 =
10.	17 + 2 =	30.	7 + 6 =
11.	13 + 5 =	31.	7 + 4 =
12.	14 + 4 =	32.	7 + 9 =
13.	16 + 3 =	33.	7 + 7 =
14.	11 + 7 =	34.	8 + 6 =
15.	9 + 2 =	35.	6 + 9 =
16.	9 + 9 =	36.	8 + 5 =
17.	6 + 9 =	37.	4 + 7 =
18.	8 + 9 =	38.	3 + 9 =
19.	7 + 8 =	39.	8 + 6 =
20.	8 + 8 =	40.	9 + 4 =

EUREKA MATH®

Lesson 14:   Use math drawings to represent subtraction with up to two decompositions, relate drawings to the algorithm, and use addition to explain why the subtraction method works.

© 2015 Great Minds®. eureka-math.org

173

Name _____   Date _____

1.	10 + 7 =	21.	5 + 8 =
2.	9 + 10 =	22.	6 + 7 =
3.	2 + 10 =	23.	____ + 4 = 12
4.	10 + 5 =	24.	____ + 7 = 13
5.	11 + 3 =	25.	6 + ____ = 14
6.	12 + 4 =	26.	7 + ____ = 14
7.	16 + 3 =	27.	____ = 9 + 8
8.	15 + ____ = 19	28.	____ = 7 + 5
9.	18 + ____ = 20	29.	____ = 4 + 8
10.	13 + 5 =	30.	3 + 9 =
11.	____ = 4 + 13	31.	6 + 7 =
12.	____ = 6 + 12	32.	8 + ____ =13
13.	____ = 14 + 6	33.	____ = 7 + 9
14.	9 + 3 =	34.	6 + 6 =
15.	7 + 9 =	35.	____ = 7 + 5
16.	____ + 4 = 11	36.	____ = 4 + 8
17.	____ + 6 = 13	37.	15 = 7 + ____
18.	____ + 5 = 12	38.	18 = ____ + 9
19.	8 + 8 =	39.	16 = ____ + 7
20.	6 + 9 =	40.	19 = 9 + ____

Name _____   Date _____

1.	$15 - 5 =$	21.	$15 - 7 =$
2.	$16 - 6 =$	22.	$18 - 9 =$
3.	$17 - 10 =$	23.	$16 - 8 =$
4.	$12 - 10 =$	24.	$15 - 6 =$
5.	$13 - 3 =$	25.	$17 - 8 =$
6.	$11 - 10 =$	26.	$14 - 6 =$
7.	$19 - 9 =$	27.	$16 - 9 =$
8.	$20 - 10 =$	28.	$13 - 8 =$
9.	$14 - 4 =$	29.	$12 - 5 =$
10.	$18 - 11 =$	30.	$11 - 2 =$
11.	$11 - 2 =$	31.	$11 - 3 =$
12.	$12 - 3 =$	32.	$13 - 8 =$
13.	$14 - 2 =$	33.	$16 - 7 =$
14.	$13 - 4 =$	34.	$12 - 7 =$
15.	$11 - 3 =$	35.	$16 - 3 =$
16.	$12 - 4 =$	36.	$19 - 14 =$
17.	$13 - 2 =$	37.	$17 - 4 =$
18.	$14 - 5 =$	38.	$18 - 16 =$
19.	$11 - 4 =$	39.	$15 - 11 =$
20.	$12 - 5 =$	40.	$20 - 16 =$

EUREKA MATH®

Lesson 14:   Use math drawings to represent subtraction with up to two decompositions, relate drawings to the algorithm, and use addition to explain why the subtraction method works.
© 2015 Great Minds®. eureka-math.org

Name _____   Date _____

1.	12 – 2 =	21.	13 – 6 =
2.	15 – 10 =	22.	15 – 9 =
3.	17 – 11 =	23.	18 – 7 =
4.	12 – 10 =	24.	14 – 8 =
5.	18 – 12 =	25.	17 – 9 =
6.	16 – 13 =	26.	12 – 9 =
7.	19 – 9 =	27.	13 – 8 =
8.	20 – 10 =	28.	15 – 7 =
9.	14 – 12 =	29.	16 – 8 =
10.	13 – 3 =	30.	14 – 7 =
11.	_____ = 11 – 2	31.	13 – 9 =
12.	_____ = 13 – 2	32.	17 – 8 =
13.	_____ = 12 – 3	33.	16 – 7 =
14.	_____ = 11 – 4	34.	_____ = 13 – 5
15.	_____ = 13 – 4	35.	_____ = 15 – 8
16.	_____ = 14 – 4	36.	_____ = 18 – 9
17.	_____ = 11 – 3	37.	_____ = 20 – 6
18.	15 – 6 =	38.	_____ = 20 – 18
19.	16 – 8 =	39.	_____ = 20 – 3
20.	12 – 5 =	40.	_____ = 20 – 11

Name _____   Date _____

1.	12 + 2 =	21.	13 – 7 =
2.	14 + 5 =	22.	11 – 8 =
3.	18 + 2 =	23.	16 – 8 =
4.	11 + 7 =	24.	12 + 6 =
5.	9 + 6 =	25.	13 + 2 =
6.	7 + 8 =	26.	9 + 11 =
7.	4 + 7 =	27.	6 + 8 =
8.	13 – 6 =	28.	7 + 9 =
9.	12 – 8 =	29.	5 + 7 =
10.	17 – 9 =	30.	13 – 7 =
11.	14 – 6 =	31.	15 – 8 =
12.	16 – 7 =	32.	11 – 9 =
13.	8 + 8 =	33.	12 – 3 =
14.	7 + 6 =	34.	14 – 5 =
15.	4 + 9 =	35.	20 – 12 =
16.	5 + 7 =	36.	8 + 5 =
17.	6 + 5 =	37.	7 + 4 =
18.	13 – 8 =	38.	7 + 8 =
19.	16 – 9 =	39.	4 + 9 =
20.	14 – 8 =	40.	9 + 11 =

A

Number Correct: _____

Subtraction from Teens

1.	11 – 10 =	
2.	12 – 10 =	
3.	13 – 10 =	
4.	19 – 10 =	
5.	11 – 1 =	
6.	12 – 2 =	
7.	13 – 3 =	
8.	17 – 7 =	
9.	11 – 2 =	
10.	11 – 3 =	
11.	11 – 4 =	
12.	11 – 8 =	
13.	18 – 8 =	
14.	13 – 4 =	
15.	13 – 5 =	
16.	13 – 6 =	
17.	13 – 8 =	
18.	16 – 6 =	
19.	12 – 3 =	
20.	12 – 4 =	
21.	12 – 5 =	
22.	12 – 9 =	

23.	19 – 9 =	
24.	15 – 6 =	
25.	15 – 7 =	
26.	15 – 9 =	
27.	20 – 10 =	
28.	14 – 5 =	
29.	14 – 6 =	
30.	14 – 7 =	
31.	14 – 9 =	
32.	15 – 5 =	
33.	17 – 8 =	
34.	17 – 9 =	
35.	18 – 8 =	
36.	16 – 7 =	
37.	16 – 8 =	
38.	16 – 9 =	
39.	17 – 10 =	
40.	12 – 8 =	
41.	18 – 9 =	
42.	11 – 9 =	
43.	15 – 8 =	
44.	13 – 7 =	

EUREKA MATH®

Lesson 16: Subtract from multiples of 100 and from numbers with zero in the tens place.

© 2015 Great Minds®. eureka-math.org

B

Number Correct: _____

Improvement: _____

Subtraction from Teens

1.	11 – 1 =	
2.	12 – 2 =	
3.	13 – 3 =	
4.	18 – 8 =	
5.	11 – 10 =	
6.	12 – 10 =	
7.	13 – 10 =	
8.	18 – 10 =	
9.	11 – 2 =	
10.	11 – 3 =	
11.	11 – 4 =	
12.	11 – 7 =	
13.	19 – 9 =	
14.	12 – 3 =	
15.	12 – 4 =	
16.	12 – 5 =	
17.	12 – 8 =	
18.	17 – 7 =	
19.	13 – 4 =	
20.	13 – 5 =	
21.	13 – 6 =	
22.	13 – 9 =	

23.	16 – 6 =	
24.	14 – 5 =	
25.	14 – 6 =	
26.	14 – 7 =	
27.	14 – 9 =	
28.	20 – 10 =	
29.	15 – 6 =	
30.	15 – 7 =	
31.	15 – 9 =	
32.	14 – 4 =	
33.	16 – 7 =	
34.	16 – 8 =	
35.	16 – 9 =	
36.	20 – 10 =	
37.	17 – 8 =	
38.	17 – 9 =	
39.	16 – 10 =	
40.	18 – 9 =	
41.	12 – 9 =	
42.	13 – 7 =	
43.	11 – 8 =	
44.	15 – 8 =	

EUREKA MATH®

Lesson 16:   Subtract from multiples of 100 and from numbers with zero in the tens place.

185

© 2015 Great Minds®. eureka-math.org

A

Number Correct: _____

Subtract Crossing the Ten

1.	10 – 1 =	
2.	10 – 2 =	
3.	20 – 2 =	
4.	40 – 2 =	
5.	10 – 2 =	
6.	11 – 2 =	
7.	21 – 2 =	
8.	51 – 2 =	
9.	10 – 3 =	
10.	11 – 3 =	
11.	21 – 3 =	
12.	61 – 3 =	
13.	10 – 4 =	
14.	11 – 4 =	
15.	21 – 4 =	
16.	71 – 4 =	
17.	10 – 5 =	
18.	11 – 5 =	
19.	21 – 5 =	
20.	81 – 5 =	
21.	10 – 6 =	
22.	11 – 6 =	

23.	21 – 6 =	
24.	91 – 6 =	
25.	10 – 7 =	
26.	11 – 7 =	
27.	31 – 7 =	
28.	10 – 8 =	
29.	11 – 8 =	
30.	41 – 8 =	
31.	10 – 9 =	
32.	11 – 9 =	
33.	51 – 9 =	
34.	12 – 3 =	
35.	82 – 3 =	
36.	13 – 5 =	
37.	73 – 5 =	
38.	14 – 6 =	
39.	84 – 6 =	
40.	15 – 8 =	
41.	95 – 8 =	
42.	16 – 7 =	
43.	46 – 7 =	
44.	68 – 9 =	

EUREKA MATH®

Lesson 17:   Subtract from multiples of 100 and from numbers with zero in the tens place.

187

B

Number Correct: _____

Improvement: _____

Subtract Crossing the Ten

1.	10 – 2 =	
2.	20 – 2 =	
3.	30 – 2 =	
4.	50 – 2 =	
5.	10 – 2 =	
6.	11 – 2 =	
7.	21 – 2 =	
8.	61 – 2 =	
9.	10 – 3 =	
10.	11 – 3 =	
11.	21 – 3 =	
12.	71 – 3 =	
13.	10 – 4 =	
14.	11 – 4 =	
15.	21 – 4 =	
16.	81 – 4 =	
17.	10 – 5 =	
18.	11 – 5 =	
19.	21 – 5 =	
20.	91 – 5 =	
21.	10 – 6 =	
22.	11 – 6 =	

23.	21 – 6 =	
24.	41 – 6 =	
25.	10 – 7 =	
26.	11 – 7 =	
27.	51 – 7 =	
28.	10 – 8 =	
29.	11 – 8 =	
30.	61 – 8 =	
31.	10 – 9 =	
32.	11 – 9 =	
33.	31 – 9 =	
34.	12 – 3 =	
35.	92 – 3 =	
36.	13 – 5 =	
37.	43 – 5 =	
38.	14 – 6 =	
39.	64 – 6 =	
40.	15 – 8 =	
41.	85 – 8 =	
42.	16 – 7 =	
43.	76 – 7 =	
44.	58 – 9 =	

 EUREKA MATH®

Lesson 17: Subtract from multiples of 100 and from numbers with zero in the tens place.

189

© 2015 Great Minds®. eureka-math.org

Credits

Great Minds® has made every effort to obtain permission for the reprinting of all copyrighted material. If any owner of copyrighted material is not acknowledged herein, please contact Great Minds for proper acknowledgment in all future editions and reprints of this module.